DIFFRACTION FROM ROUGH SURFACES AND DYNAMIC GROWTH FRONTS

DIFFRACTION FROM ROUGH SURFACES AND DYNAMIC GROWTH FRONTS

H.-N. Yang
G.-C. Wang
T.-M. Lu

Centre for Integrated Electronics,
Rensselaer Polytechnic Institute,
Troy, NY 12180,
USA

World Scientific
Singapore • New Jersey • London • Hong Kong

Published by

World Scientific Publishing Co. Pte. Ltd.

P O Box 128, Farrer Road, Singapore 9128

USA office: Suite 1B, 1060 Main Street, River Edge, NJ 07661

UK office: 73 Lynton Mead, Totteridge, London N20 8DH

DIFFRACTION FROM ROUGH SURFACES AND DYNAMIC GROWTH FRONTS

ISBN 981-02-1536-3

Printed in Singapore.

PREFACE

The growth of thin films is not only a subject of great interest from the practical point of view but is also of fundamental scientific interest. It is probably not exaggerating to say that the technological foundations of most high-tech industries are based primarily on the development and utilization of various kinds of thin films. Despite decades of study, many important physical processes associated with film growth are still not well understood.

A particularly interesting subject in thin film growth is the morphology of the surface growth fronts which directly controls many physical and chemical properties of the film eventually grown. Growth processes are inherently non-equilibrium processes. A standard statistical mechanics approach to a non-equilibrium process has not been developed so far, and cannot be used to described the growth dynamics and therefore the complex morphology of the growth fronts.

In the simpler cases, with sufficient surface mobility, the film growth proceeds in either a layer-by-layer manner, or in the form of three-dimensional islands, or a combination of both. These processes have been described in the past based on two-dimensional or three-dimensional nucleation and growth theories. However, very often the surface (or the vacuum-solid interface) is rough due to the lack of surface mobility and fluctuations in the deposition rate and/or substrate temperature. Such dynamic roughening is usually stronger than that due to thermal fluctuation alone. Only very recently the theoretical understanding of the dynamical case has been pursued based on a very interesting dynamic scaling approach. To date, this approach represents the best and perhaps the most promising way to describe the interface growth problem, even though very little experimental measurement has been made to confirm the validity of this approach.

Because of the aforementioned progress made in theoretical description of the interface growth problem, the authors expect an increase in experimental study in the area of accurate growth front measurements, particularly for rough growth fronts. The most direct way to measure the relevant growth parameters such as the interface width and the surface roughness parameter is perhaps the use of the surface imaging techniques such as Scanning Tunneling Microscopy (STM). However, in addition to the experimental issues of the compatibility of the STM technique in the growth environment one also has to worry about the interpretation of the data. One notable issue is the finite size of the tip used in the measurement, which may obscure the true surface profile in an STM scan, particularly in the local region. Nevertheless STM will likely be one of the popular techniques used in surface roughness study.

A more powerful class of experimental approaches is perhaps the use of diffraction techniques. Examples are: high-resolution low-energy electron diffraction (HRLEED), reflection high-energy electron diffraction (RHEED), glancing X-ray diffraction, atom diffraction, neutron diffraction, and diffuse light scattering. In diffraction one measures the structure factors which contain quantitative information on the statistical properties of the surface. Structure factors are basically Fourier transforms of the real space images. Structure factors (in reciprocal space) are a fundamental measure of the dynamic behavior of the growing interface and can be directly obtained through the intensity distribution of the diffracted beams.

This monograph has been designed for experimentalists who wish to study the dynamics of thin film growth using diffraction techniques, and also for theorists who would like to learn the dynamic behavior of the interface growth in Fourier space. The monograph is aimed at readers who have very little prior knowledge of diffraction. Graduate and advanced undergraduate students can pick up the subject matter in the monograph with little difficulty. Yet the monograph would quickly bring the readers to forefront research in the area of the dynamics of interface growth. After the introduction of the elementary theory of diffraction in Chapter I, we discuss in Chapter II the various parameters and correlation functions that are essential in describing a rough surface. Examples are the roughness parameter, vertical correlation length (interface width), lateral correlation length, height-height correlation function, and the height difference function. In Chapter III we calculate the diffraction structure factor for both rough crystalline and non-crystalline surfaces. The differences between a crystalline and a non-crystalline surface are detailed. The methods of extracting the interface width, the lateral correlation length, and the roughness parameter from the diffraction structure factor are outlined. In Chapter IV, we present the basic physical concepts underlying the scaling hypothesis. The dynamic scaling properties of the height-height correlation function, the height difference function, and the structure factor are described. In Chapter V, the structure factor from a dynamic growth front is derived. Methods of extracting various growth exponents from the structure factor at different diffraction conditions are given in detail. An example of a quantitative measurement of the dynamic growth front of an epitaxial system is described. A particular type of rough surfaces having a diverging interface width associated with an equilibrium surface roughening transition is discussed in Chapter VI. A comparison of the diffraction characteristics from divergent and non-divergent interface is summarized in Chapter VI. We also give a brief review and summary at the end of each chapter. In this monograph we do not cover the conventional equilibrium growth problem, neither the layer-by-layer growth nor the three dimensional island growth.

Two of us (TML and GCW) would like to take this opportunity to acknowledge our mentor and collaborator Professor M. G. Lagally for his inspiration and continuous encouragement throughout our careers. We thank the numerous contributors of the work collected in this monograph. Particularly we would like to thank Drs. J. M. Pimbley, M. Henzler, P. I. Cohen, and S. K. Sinha for their contributions to diffraction theories collected in this monograph. We are deeply grateful to Professor F. Family for his many instructive discussions regarding the scaling dynamics during growth. The support of an NSF grant to perform part of the work over the last several years is deeply appreciated. Finally, it is a pleasure to thank Prof. J. S. Levinger for proofreading this manuscript.

H.-N. Yang, Dr.

G.-C. Wang, Professor

T.-M. Lu, Professor and Chair

Department of Physics
Rensselaer Polytechnic Institute
Troy, NY

CONTENTS

Chapter I FUNDAMENTALS OF DIFFRACTION

In this chapter we first give an elementary treatment of the scattering of a wave from an atom and then the interference of the scattered waves from a two-dimensional array of atoms. The concepts of momentum transfer parallel and perpendicular to the surface, which play a very important role in the diffraction of waves from surfaces, are introduced. The two-dimensional reciprocal space structure is also introduced in an elementary manner.

§I.1 Scattering and Diffraction Concepts

§I.1.1 *Scattering of waves from a single atom*

We consider a plane wave in the form of $\exp(i\mathbf{k}_0 \cdot \boldsymbol{\rho})$ incident upon an atom. $\mathbf{k}_0$ is the incident wave vector with a magnitude of $2\pi/\lambda$, where λ is the wavelength. $\boldsymbol{\rho}$ is the position of the atom with respect to a reference point $\boldsymbol{\rho} = 0$. The argument $i\mathbf{k}_0 \cdot \boldsymbol{\rho}$ describes the phase of the wave at the position of the atom. The wave can be an electron, atom, or photon (X-ray or light). Figure 1.1 shows a schematic of the diffraction geometry.

The wave interacts with the atom electromagnetically. The scattered spherical wave is received by a detector at point P a distance $\mathbf{R}$ away from the origin. The scattered wave has a wave vector of $\mathbf{k}_S$. The distance between the atom and the detector is $\mathbf{r}$. We consider only elastic scattering so that the wavevector of the out-going wave, $\mathbf{k}_S$, has the same magnitude as the incident wave vector $\mathbf{k}_0$, i.e., $|\mathbf{k}_S| = |\mathbf{k}_0|$, or $k_S = k_0$. The scattered wave has a phase of $ik_S r$ at point P with reference to the atom. That is, the wave has gained a phase of $ik_S r$ after it propagates from the position of the atom to the detector at point P. Therefore the total phase of the wave at the detector at point P is $i(\mathbf{k}_0 \cdot \boldsymbol{\rho} + k_S r)$. This phase gives a phase factor of $\exp[i(\mathbf{k}_0 \cdot \boldsymbol{\rho} + k_S r)]$. The total amplitude of the wave, $A(\mathbf{k}_0, \mathbf{k}_S)$, at point P can therefore be written as,[1.1]

$$A(\mathbf{k}_0, \mathbf{k}_S) \sim \frac{1}{r} e^{i(\mathbf{k}_0 \cdot \boldsymbol{\rho} + k_S r)} . \qquad (1.1)$$

The factor 1/r which appears in front of the phase factor is due to the spreading of the spherical wave scattered from the atom. We have ignored the atomic scattering factor which is a slowly varying function of the magnitude of the wave vector and the relative angle between the incident and outgoing waves. In a diffraction experiment, the spread of the incident wave vector and the angular spread of a diffraction spot are normally very small so that the atomic structure factor can be treated as a constant over the diffraction spot and does not play a significant role in the interpretation of the data.

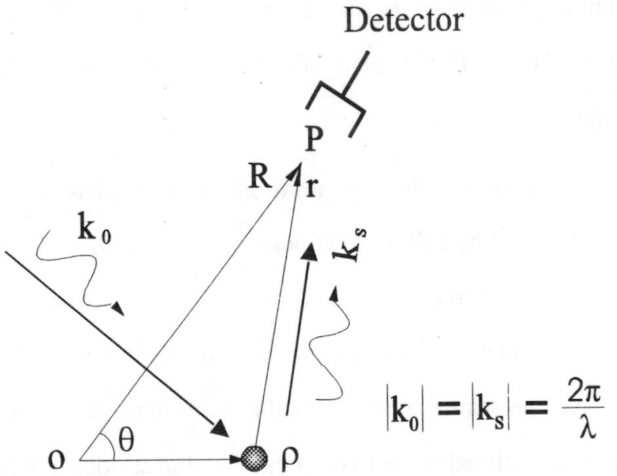

Fig. 1.1 Schematic of the geometry for the wave scattered by an atom and the detection of the scattered wave.

Normally the detector is placed very far away from the scatterer. The origin ρ=0 is assumed to be near the vicinity of the atom. In the diffraction geometry shown in Fig. 1.1, we have R >> ρ. Then $r \approx R - \rho\cos\theta$. The phase in Eq. (1.1) can be approximately written as

$$i(\mathbf{k_0}\cdot\boldsymbol{\rho} + k_S r) \approx i(\mathbf{k_0}\cdot\boldsymbol{\rho} - k_S\rho\cos\theta) + ik_S R\,.$$

Since the direction of **R** and **r** are approximately the same looking from the detector, the factor in the parentheses of the right hand side of the above equation can be written as

$$i(\mathbf{k_0}\cdot\boldsymbol{\rho} - k_S\rho\cos\theta) \approx i[(\mathbf{k_0}-\mathbf{k_S})\cdot\boldsymbol{\rho}].$$

Now the amplitude of the wave given by Eq. (1.1) can be written in a simple form:

$$A(\,\mathbf{k_0}, \mathbf{k_S}\,) \approx \frac{e^{ik_S R}}{R}\, e^{\,i(\mathbf{k_0}-\mathbf{k_S})\cdot\boldsymbol{\rho}} \propto e^{-i\mathbf{k}\cdot\boldsymbol{\rho}}, \qquad (1.2)$$

where $\mathbf{k} = \mathbf{k_S} - \mathbf{k_0}$ is defined as the momentum transfer in the diffraction. As we shall see later the momentum transfer plays a very important role in diffraction. In the last step, we have ignore the constant attenuation factor 1/R, since we are not interested in the absolute intensity of the diffraction wave. (Recall that R is the distance between the origin to the detector.) We have also omitted the constant phase factor $\exp(ik_S R)$. Even in the case of the diffraction from multiple atoms, the phase $ik_S R$ remains a constant.

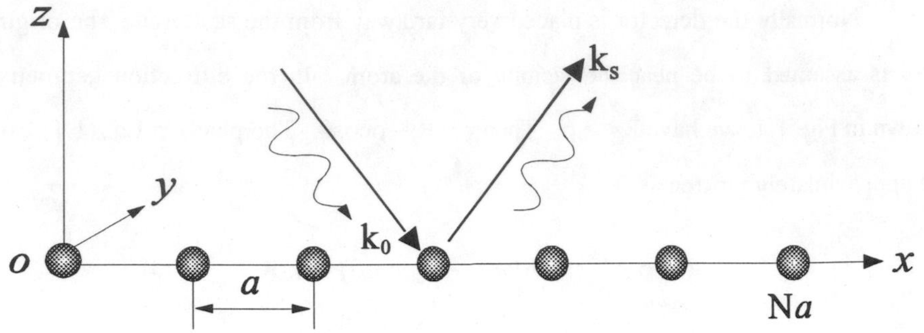

Fig. 1.2 Diffraction geometry from a one-dimensional "crystal".

§I.1.2 *Diffraction from an array of atoms*

Let us now consider the diffraction of a wave from an array of N atoms with a lattice constant $\boldsymbol{a}$. Figure 1.2 shows a schematic of the diffraction geometry. Na is the size of the one-dimensional "crystal".

Note that although the "crystal" is one dimension, the diffraction itself is not restricted to one-dimensional space. In fact, the diffraction can be three-dimensional. For simplicity, we shall consider the diffraction to be confined to the xz plane. That is, all the diffraction wave vectors are in the xz plane and the array of atoms are aligned in the x–direction at $z = 0$.

We choose the origin $\boldsymbol{\rho}_1 = 0$ at the position of the first atom in the row. The nth atom is located at $\boldsymbol{\rho}_n = n\boldsymbol{a}$. The total scattering amplitude received by the detector at point P is the sum of the individual scattering amplitudes from the array of atoms:

$$A(\mathbf{k}) = e^{-i\mathbf{k}\cdot\boldsymbol{\rho}_1} + e^{-i\mathbf{k}\cdot\boldsymbol{\rho}_2} + \ldots\ldots + e^{-i\mathbf{k}\cdot\boldsymbol{\rho}_N}$$

$$= \sum_{n=0}^{N-1} e^{-i\,n\mathbf{k}\cdot\boldsymbol{a}} = \frac{1 - e^{-i\,N\mathbf{k}\cdot\boldsymbol{a}}}{1 - e^{-i\,\mathbf{k}\cdot\boldsymbol{a}}} . \qquad (1.3)$$

We evaluate the sum of a geometrical series to find the last expression.

The total intensity of the diffracted wave is the product of Eq. (1.3) with its own complex conjugate and has the following form:

$$S(\mathbf{k}) = A^*(\mathbf{k})\,A(\mathbf{k}) = \frac{\sin^2(N\mathbf{k}\cdot\boldsymbol{a}/2)}{\sin^2(\mathbf{k}\cdot\boldsymbol{a}/2)} . \qquad (1.4)$$

A more general way of treating this diffraction problem is to calculate the scattering amplitude as a Fourier transform of the density function (or source function), $D(\boldsymbol{\rho})$, for the corresponding scatterers,

$$A(\mathbf{k}) = \int D(\boldsymbol{\rho})\, e^{-i\mathbf{k}\cdot\boldsymbol{\rho}}\, d^3\rho . \qquad (1.5)$$

The Fourier transform is between the real space (scatterer or source) $\boldsymbol{\rho}$ and the momentum transfer space $\mathbf{k}$. In our case of the one-dimensional array, the scatterers are atoms and the density function can be represented by

$$D(\boldsymbol{\rho}) = \sum_{n=0}^{N-1} \delta(\,\boldsymbol{\rho} - n\boldsymbol{a}\,).$$

δ is the Dirac delta function having the properties:

$$\delta(\,x - x_0\,) = \begin{cases} \infty & \text{for } x = x_0, \\ 0 & \text{for } x \neq x_0, \end{cases}$$

and

$$\int_{-\infty}^{+\infty} dx\ \delta(x - x_0)\ f(x) = f(x_0)\ .$$

The diffraction amplitude can therefore be written as

$$A(\mathbf{k}) = \sum_{n=0}^{N-1} \int \delta(\rho - n\boldsymbol{a})\ e^{-i\mathbf{k}\cdot\rho}\ d^3\rho$$

$$= \sum_{n=0}^{N-1} \int \delta(x - na)\ \delta(y)\ \delta(z)\ e^{-i\mathbf{k}\cdot\rho}\ dxdydz$$

$$= \sum_{n=0}^{N-1} e^{-i\,n\mathbf{k}\cdot\boldsymbol{a}}\ , \qquad (1.6)$$

where we have defined $\rho = (x, y, z)$. We see that Eq. (1.6) is identical to (1.3). The later approach (which leads to Eq. (1.6)) is more general and will be used again later to calculate the diffraction intensity from a source with a different charge distribution.

As an example, we can use Eq. (1.5) to calculate the diffraction from a two-dimensional array with lattice constants, $\boldsymbol{a}$ and $\boldsymbol{b}$, along x and y directions, respectively. Figure 1.3 shows a schematic of the diffraction geometry. $N_x a$ and $N_y b$ are the sizes of the two-dimensional "crystal" in the two directions. The scattering density is then given by

$$D(\rho) = \sum_{n=0}^{N_x-1} \sum_{m=0}^{N_y-1} \delta(\rho - n\boldsymbol{a} - m\boldsymbol{b})\ .$$

The diffraction amplitude can be calculated using Eq. (1.5) as

$$A(\mathbf{k}) = \sum_{n=0}^{N_x-1} e^{-i\,n\mathbf{k}\cdot\boldsymbol{a}} \times \sum_{m=0}^{N_y-1} e^{-i\,m\mathbf{k}\cdot\boldsymbol{b}} .$$

The total intensity of the diffracted wave is represented as the product of the intensity functions from two 1D arrays shown in Eq. (1.4):

$$S(\mathbf{k}) = A^*(\mathbf{k})\,A(\mathbf{k}) = \frac{\sin^2(N_x\mathbf{k}\cdot\boldsymbol{a}/2)}{\sin^2(\mathbf{k}\cdot\boldsymbol{a}/2)} \times \frac{\sin^2(N_y\mathbf{k}\cdot\boldsymbol{b}/2)}{\sin^2(\mathbf{k}\cdot\boldsymbol{b}/2)} . \tag{1.7}$$

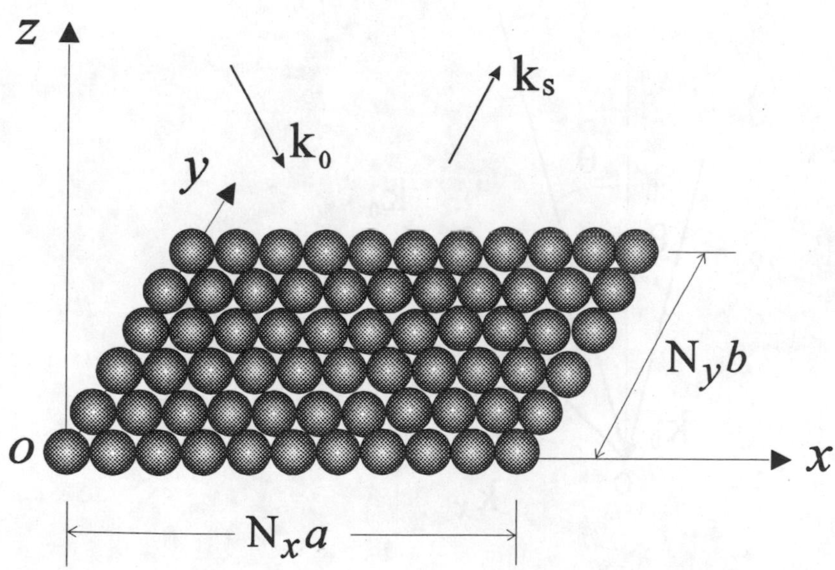

Fig. 1.3 Diffraction geometry from a two-dimensional "crystal".

§I.2 Reciprocal Space Characteristics

§I.2.1 *Diffraction pattern from an array of atoms*

The most relevant parameter in the intensity function given by Eq. (1.4) is the momentum transfer along the direction of the array of atoms, i.e., the x-direction shown in Fig. 1.2. We can write $\mathbf{k}\cdot\boldsymbol{a} = k_x a$, where k_x is defined to be the momentum transfer parallel to the *x*-direction and is shown in Fig. 1.4.

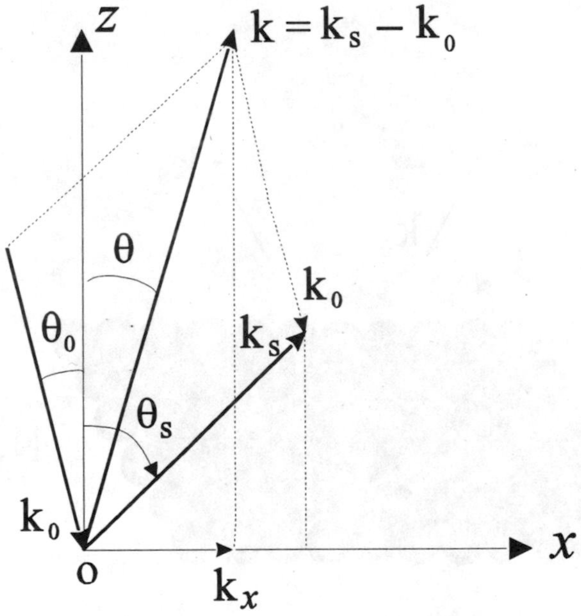

Fig. 1.4 Wavevector diagram of diffraction geometry.

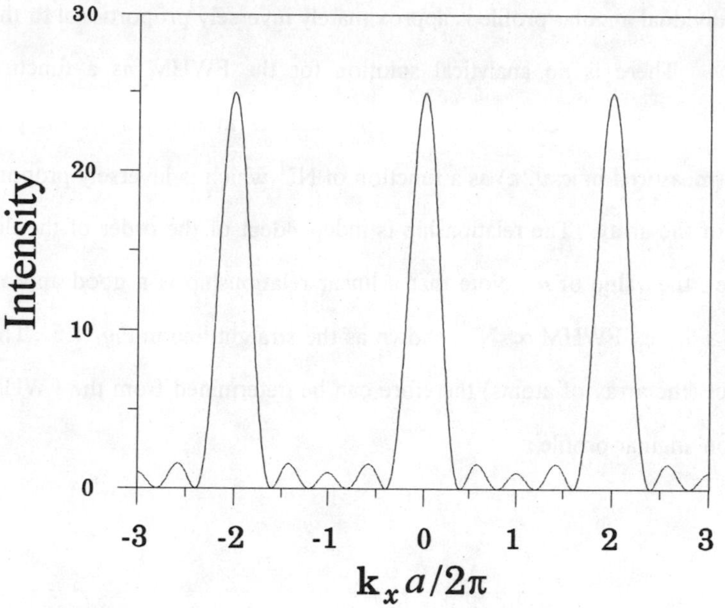

Fig. 1.5 Diffraction intensity distribution from the one-dimensional crystal for N = 5.

According to Fig. 1.4, the momentum transfer parallel to the *x*-direction can be written as

$$k_x = k_s \sin\theta_s - k_0 \sin\theta_0 , \quad (1.8)$$

where θ_0 and θ_s are the incident and outgoing angles of the wave with respect to the normal, or the *z*-direction. For a fixed incident angle, k_x is a function of the wavelength and the diffracted angle. It is convenient to use $k_x a$ as a variable to describe the diffraction intensity. At $k_x a = 2n\pi$, where *n* is an integer, the intensity has its maximum value and is equal to N^2. In Fig. 1.5, the diffraction intensity is plotted as a function of $k_x a/2\pi$ for N=5. The full-width-at-half-maximum (FWHM) of the individual angular profile is approximately inversely proportional to the size of the array. There is no analytical solution for the FWHM as a function of N. Obtained by numerical calculation, the open squares in Fig. 1.6 are the plot of the FWHM (measured in $k_x a/2\pi$) as a function of N^{-1} which is inversely proportional to the size of the array. The relationship is independent of the order of the diffraction beam, i.e., the value of *n*. Note that a linear relationship is a good approximation when $N \geq 5$, i.e., FWHM $\propto N^{-1}$, shown as the straight line in Fig. 1.6. The size of the object (the array of atoms) therefore can be determined from the FWHM of the diffraction angular profile.

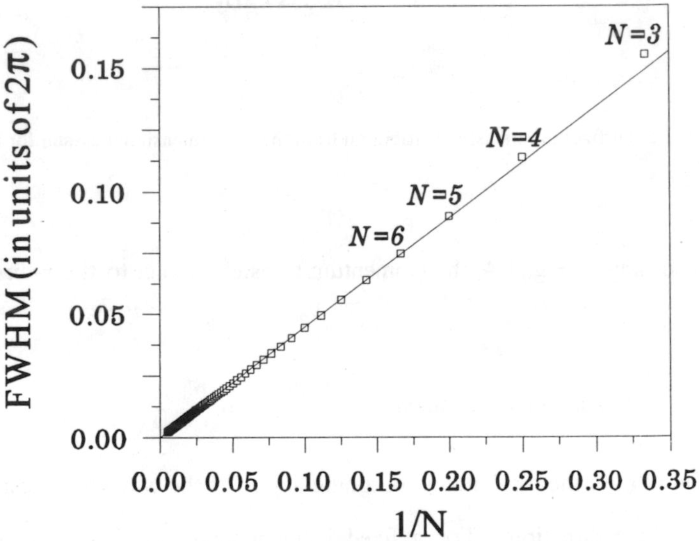

Fig. 1.6 The FWHM of the diffraction intensity distribution as a function of N^{-1}. The FWHM is measured in $k_x a/2\pi$ as shown in Fig. 1.5.

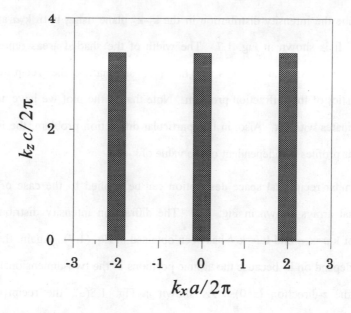

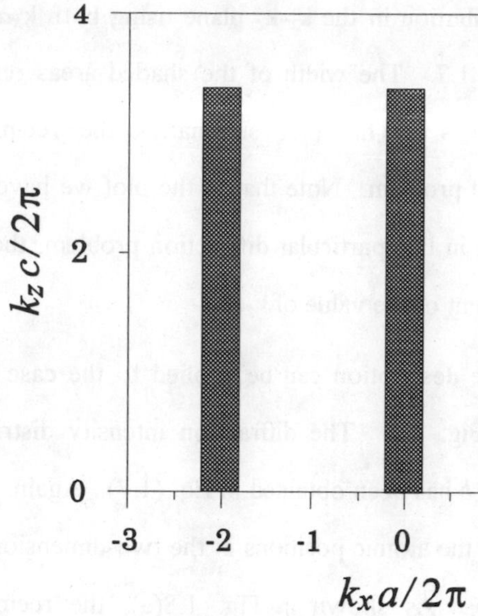

Fig. 1.7 Reciprocal-space characteristics of the diffraction intensity distribution from the one-dimensional crystal. The FWHM of the angular profiles determines the size of the shaded area.

The intensity given by Eq. (1.4) is independent of k_z, the momentum transfer in the *z*-direction. This is because the position of the array of atoms is fixed in the *z*-direction (*z*=0). If different atoms are located at different positions with respect to the *z*-direction, then the intensity in general may depend on k_z. As we shall see later, one of the most important defects that occurs in surfaces is the surface atomic steps in which layers of atoms are situated at different *z*-positions. In this case, k_z can be an important parameter to describe surfaces with defects. Experimentally, one can vary k_x without changing the value of k_z, and vice versa. They are independent variables. For convenience, we shall use $k_z c$, where *c* is the step height, instead of k_z, as the variable. In our present case of the diffraction from an array of atoms, one can describe the intensity distribution in the k_x-k_z plane using both $k_x a$ and $k_z c$ as variables. It is shown in Fig. 1.7. The width of the shaded areas represents the FWHM of the angular profiles. This plot summaries the reciprocal space characteristics of the diffraction problem. Note that in the plot we have normalized the coordinates with 2π. Also, in this particular diffraction problem, the FWHM of the angular profiles is independent of the value of $k_z c$.

Similar reciprocal space description can be applied to the case of the two-dimensional arrays shown in Fig. 1.3. The diffraction intensity distribution as a function of $\mathbf{k}\cdot\boldsymbol{a}=k_x a$ and $\mathbf{k}\cdot\boldsymbol{b}=k_y b$ has been obtained in Eq. (1.7). Again, the intensity does not depend on k_z because the atomic positions in the two-dimensional array are fixed in the *z*-direction (z=0). As shown in Fig. 1.8(a), the reciprocal space characteristics of the diffraction intensity is plotted in a three-dimensional k_x-k_y-k_z space using $k_x a$, $k_x b$ and $k_z c$ as variables. In Fig. 1.8(a), we have a set of rods which are parallel to the *z*-direction and are centered at the maximum intensity positions, ($k_x a = 2n\pi$, $k_y b = 2m\pi$), n, m = 0, ±1, ±2, ... The maximum intensity is given by Eq. (1.7) as $(N_x N_y)^2$. Each rod represents the region in which the intensities are higher than the half of the maximum in the corresponding angular profile. If the set of rods shown in Fig. 1.8(a) are projected onto the k_x-k_z or k_y-k_z plane, we obtain similar two-dimensional plots shown in Fig. 1.7.

We can also take a look at the cross sections of these rods in the k_x-k_y plane, as shown in Fig. 1.8(b). The k_x-k_y cross section for each rod gives approximately an elliptical shape if $N_x \neq N_y$. For $N_x = N_y$, the cross section becomes a circle-like shape. Such an ellipse-like cross section represents the diffraction spots (or image) measured by the detector as scanned across the diffraction beam. The two principal-axes of the "elliptical" spots are half of the full-width-at-half-maximum (FWHM) of the individual angular profile along the k_x and k_y directions, respectively. Similar to the one-dimensional case discussed above, the FWHM's along k_x and k_y directions, and then the two principal-axes of the "elliptical" spots, are inversely proportional to N_x and N_y, respectively. Therefore, the size of each diffraction spot is solely determined by the size of the two-dimensional "crystal" shown in Fig. 1.3.

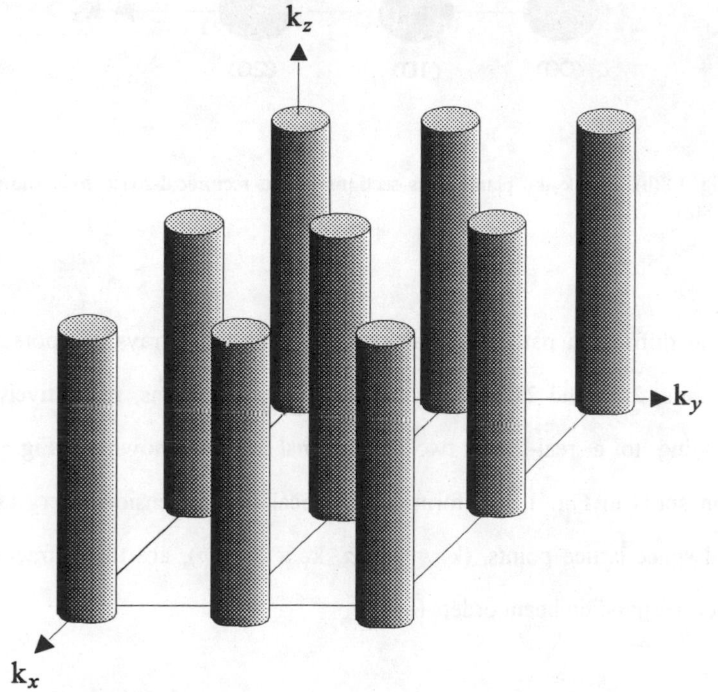

Fig. 1.8(a) Reciprocal-space characteristics of diffraction intensity distribution from the two-dimensional crystal. The FWHM of the angular profiles determines the shape and diameter of each rod.

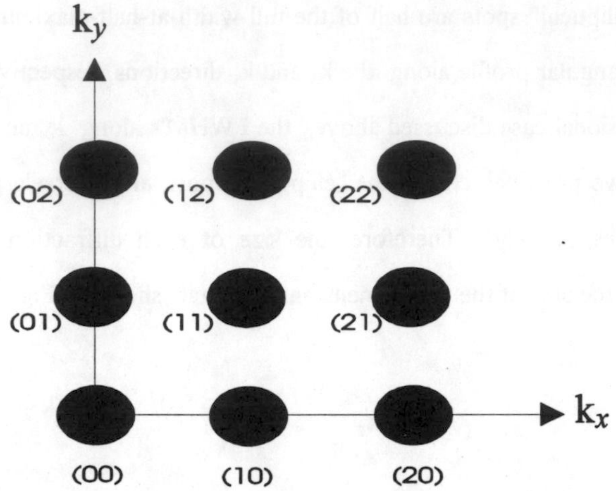

Fig. 1.8(b) The k_x-k_y plane cross sections of the reciprocal-space rods shown in Fig. 1.8(a).

The diffraction pattern contains two-dimensional arrays of spots with the periodic units, $2\pi/a$ and $2\pi/b$, along the k_x and k_y directions, respectively. Thus, corresponding to a real-space two-dimensional crystal shown in Fig. 1.3, the diffraction spots in Fig. 1.8(b) form a reciprocal two-dimensional "crystal". The reciprocal space lattice points, ($k_x = 2n\pi/a$, $k_y = 2m\pi/b$), are the diffraction spots with the corresponding beam orders (n, m).

§I.2.2 *Diffraction characteristics from a perfect crystalline surface*

We can extend the above discussion to a practical diffraction problem: the diffraction from a perfect crystalline surface. The surface as the termination (or boundary) of a crystal is drawn schematically in Fig. 1.9. The crystal has three orthogonal unit vectors, ***a***, ***b***, ***c***, along *x*, *y* and *z* directions, respectively. The size of the crystal is $(N_x a) \times (N_y b) \times (N_z c)$. The normal of the surface is chosen to be parallel to the *z*-direction.

The scattering density in this specific case is given by

$$D(\rho) = \sum_{l=0}^{-(N_z-1)} \sum_{n=0}^{N_x-1} \sum_{m=0}^{N_y-1} \delta(\rho - n\mathbf{a} - m\mathbf{b} - l\mathbf{c}).$$

The negative sign in front of (N_z-1) means that the summation is from the surface level, l=0, into the levels beneath the surface.

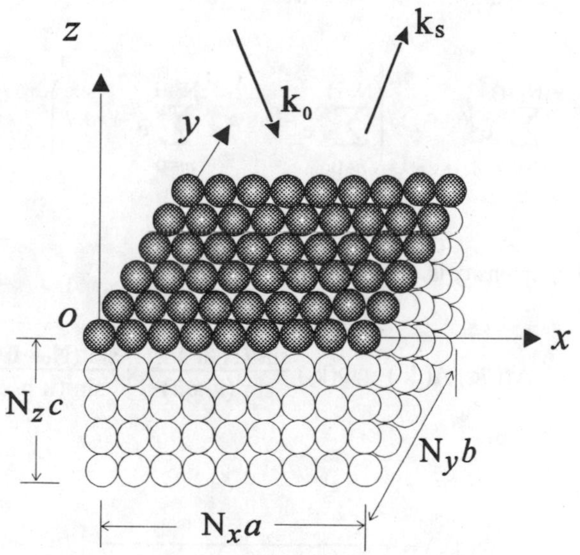

Fig. 1.9 The diffraction geometry from the surface of a three-dimensional crystal.

For most of the surface-sensitive diffraction techniques, such as the low-energy electron diffraction (LEED) technique, the incident wave cannot penetrate deeply into a crystal because of the partial scattering or reflection by atoms in the top surface layers. [1.2, 1.3] Therefore, the incident wave has an attenuated amplitude inside a crystal. The attenuation of the diffraction wave inside a crystal can be approximately described by an attenuation factor, $\exp(-\kappa|z|)$. The attenuation parameter κ is a measure of the surface penetration depth in the diffraction. For the present problem, we assume the penetration depth, $\Delta l \sim 1/\kappa \ll N_z c$. Thus, for the crystalline surface shown in Fig. 1.9, the total diffraction amplitude, Eq. (1.5), can be calculated as [1.2, 1.3]

$$A(\mathbf{k}) = \int D(\rho)\, e^{-\kappa|z|}\, e^{-i\mathbf{k}\cdot\rho}\, d^3\rho$$

$$= \sum_{l=0}^{-(N_z-1)} e^{l(\kappa - ik_z)c} \left(\sum_{n=0}^{N_x-1} e^{-i n\mathbf{k}\cdot\boldsymbol{a}} \times \sum_{m=0}^{N_y-1} e^{-i m\mathbf{k}\cdot\boldsymbol{b}} \right).$$

The corresponding intensity is

$$S(\mathbf{k}) = A^*(\mathbf{k})\, A(\mathbf{k}) = \Phi(k_z) \frac{\sin^2(N_x\mathbf{k}\cdot\mathbf{a}/2)}{\sin^2(\mathbf{k}\cdot\mathbf{a}/2)} \times \frac{\sin^2(N_y\mathbf{k}\cdot\mathbf{b}/2)}{\sin^2(\mathbf{k}\cdot\mathbf{b}/2)}, \qquad (1.7')$$

where

$$\Phi(k_z) = \frac{1}{1 + e^{-2\kappa c} - 2e^{-\kappa c}\cos(k_z c)}.$$

We can compare the diffraction intensity from a two-dimensional crystal, given by Eq. (1.7), with the diffraction intensity from a crystalline surface which has a three-dimensional crystalline structure, given by Eq. (1.7'). Except for the factor $\Phi(k_z)$, the two intensity expressions are identical. The function $\Phi(k_z)$ depends only on k_z. If the diffraction intensity is measured in the k_x–k_y plane by the detector so that k_z remains a constant, the intensity distribution can then be described by Eq. (1.7) because $\Phi(k_z)$ modifies only the absolute intensity but not the relative diffraction intensity distribution. Therefore, the reciprocal-space description introduced in **§I.2.1**, as shown in Fig. 1.8, is still valid in the present case of a perfect crystalline surface. Examples of the k_x–k_y plane measurement can be found in High-Resolution Low Energy Electron Diffraction (HRLEED),[1.4, 1.5] atomic scattering,[1.6, 1.7] X-ray [1.8, 1.9, 1.10] and Reflection High-Energy Electron Diffraction (RHEED) [1.11] in-plane scattering techniques, which can be explained as follows.

In the case of the HRLEED experiment, for example, the incident and diffraction beams are near the normal direction, i.e., $\theta_s \sim 0$ and $\theta_o \sim 0$, as shown in Fig. 1.10(a). Accordingly, the corresponding angular scan is approximately in the k_x–k_y plane because $k_z = k_s \cos\theta_s + k_o \cos\theta_o \approx \frac{4\pi}{\lambda} = \text{constant}$.

For the case of finite incident and diffraction angles, the angular scan that is along the azimuth direction (within k_x–k_y plane) can still keep k_z as a constant. Figure 1.10(b) shows the diffraction geometry for the grazing angle in-plane scattering measurement, as employed in X-ray and RHEED techniques. Using the spherical coordinates in reciprocal space (k-space), we have

$$\begin{cases} k_x = k \sin\theta \cos\phi \\ k_y = k \sin\theta \sin\phi \\ k_z = k \cos\theta \end{cases},$$

where θ is the polar angle of the vector **k** with respect to the surface normal (z-direction) and ϕ denotes the azimuth angle. For the diffraction geometry shown in 1.10(b), 2ϕ gives the azimuth angle of $\mathbf{k}_s$ and the 2ϕ–angle scan of the detector is in the k_x–k_y plane.

However, if the diffraction intensity is not measured in the k_x–k_y plane by the detector, we must consider the modification of $\Phi(k_z)$. Examples are the X-ray and RHEED specular reflection measurements, as shown in Fig. 1.10(c). For the grazing angle diffraction geometry shown in Fig. 1.10(c), we have, $\theta_o \sim \frac{\pi}{2}$, $\theta_s \sim \frac{\pi}{2}$ and

$$k_z = k_s \cos\theta_s + k_o \cos\theta_o = \frac{2\pi}{\lambda}[\sin(\frac{\pi}{2} - \theta_s) + \sin(\frac{\pi}{2} - \theta_o)] \approx \frac{2\pi}{\lambda}(\pi - \theta_s - \theta_o).$$

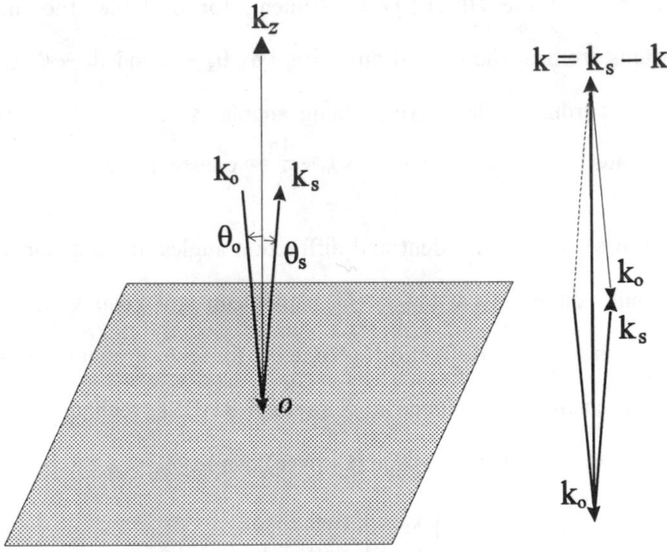

Fig. 1.10(a) Diffraction geometry for HRLEED.

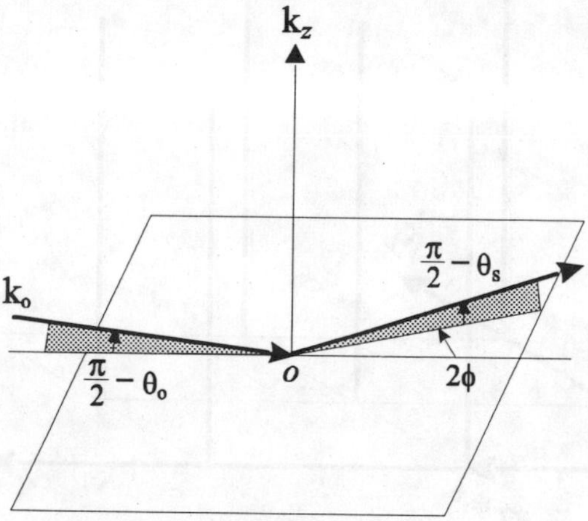

Fig. 1.10(b) Diffraction geometry for a grazing angle in-plane scattering measurement, which is usually employed in X-ray and RHEED techniques.

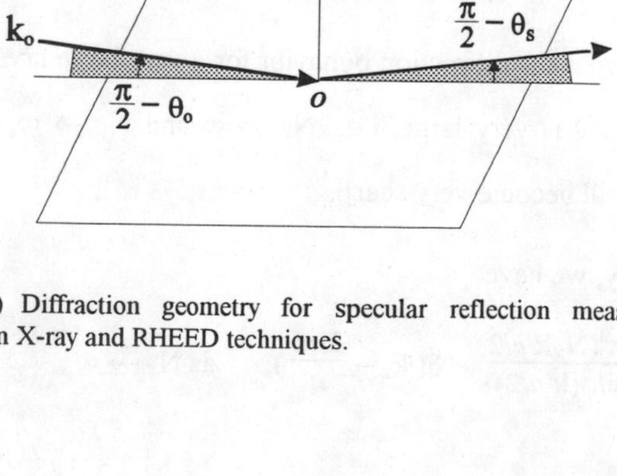

Fig. 1.10(c) Diffraction geometry for specular reflection measurement, as usually employed in X-ray and RHEED techniques.

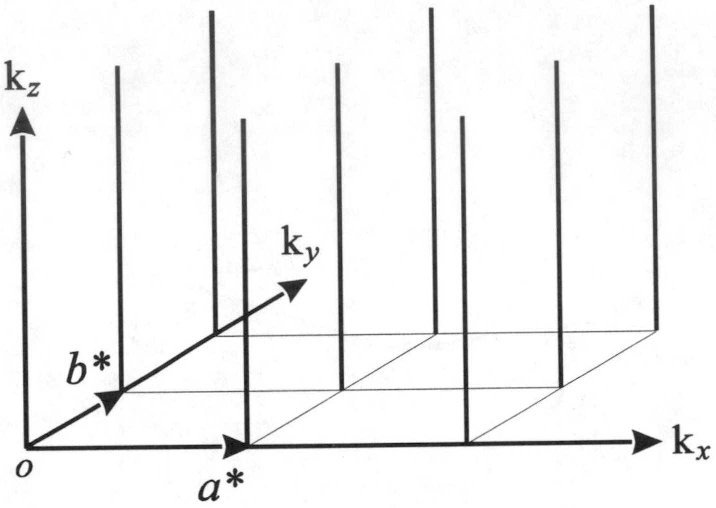

Fig. 1.11 Reciprocal-space characteristics of diffraction intensity distribution from an infinitely large and flat surface.

If the diffraction spot is scanned out of the k_x–k_y plane, one must vary either θ_s or θ_o, or both. As a result, k_z must change significantly. Therefore, the measured intensity distribution depends on the out-of-plane prefactor, $\Phi(k_z)$. Keeping this in mind, we shall not include the factor $\Phi(k_z)$ in the discussion of intensity in subsequent chapters except for special occasions.

Now let us discuss the diffraction behavior for a very large crystal. If the size of the crystal in Fig. 1.9 is very large, i.e., $N_x \to \infty$ and $N_y \to \infty$, the diffraction intensity distribution will become very sharp.

Mathematically, we have

$$\frac{\sin^2(N_x \mathbf{k}\cdot\boldsymbol{a}/2)}{\sin^2(\mathbf{k}\cdot\boldsymbol{a}/2)} \sim \delta\left(k_x - \frac{2n\pi}{a}\right), \qquad \text{as } N_x \to \infty,$$

and

$$\frac{\sin^2(N_y \mathbf{k}\cdot \boldsymbol{b}/2)}{\sin^2(\mathbf{k}\cdot \boldsymbol{b}/2)} \sim \delta(\, k_y - \frac{2m\pi}{b}\,), \qquad \text{as } N_y \to \infty\,.$$

The diffraction intensity distribution can thus be represented by a sum of δ–functions,

$$S(\,\mathbf{k}\,) \sim \sum_{n,\,m} \delta(\,\mathbf{k}_{\|} - \frac{2n\pi}{a}\mathbf{e}_x - \frac{2m\pi}{b}\mathbf{e}_y\,), \tag{1.9}$$

where $\mathbf{k}_{\|} = k_x\mathbf{e}_x + k_y\mathbf{e}_y$ is the momentum transfer parallel to the surface. Equation (1.9) describes the diffraction characteristics of an infinitely large and flat surface. The appearance of the δ–intensity is a measure of the flatness (or perfectness) of the surface in a large system. This very important property will be frequently utilized in the following chapters. Similar to Fig. 1.8, the reciprocal space characteristics in the present case can be represented by Fig. 1.11. We can see that the rods with a finite width shown in Fig. 1.8(a) have been replaced by the straight lines with a negligible width as shown in Fig. 1.11.

§I.2.3 *Two-dimensional reciprocal lattice*

As shown in Fig. 1.8(b), there is a one to one correspondence between the two-dimensional real space crystalline structure and the diffraction pattern. The diffraction pattern forms a two-dimensional periodical lattice structure in the reciprocal space. In this section, we shall formally define the reciprocal lattice through the two-dimensional real space structure.

We start from a general two-dimensional perfect lattice shown in Fig. 1.12(a). The lattice has translational symmetry with the space periodicity, a and b, along $\boldsymbol{a}$ and $\boldsymbol{b}$ directions, respectively. For the general case, the unit vectors, $\boldsymbol{a}$ and $\boldsymbol{b}$, may not be necessarily perpendicular to each other. We denote the surface normal direction by the normal vector, $\mathbf{n}$.

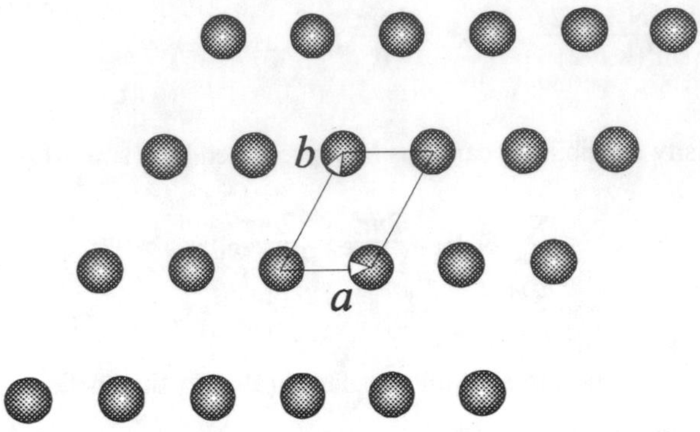

Fig. 1.12(a) A generic real-space two-dimensional lattice with the unit-mesh vectors given by (***a*** , ***b***).

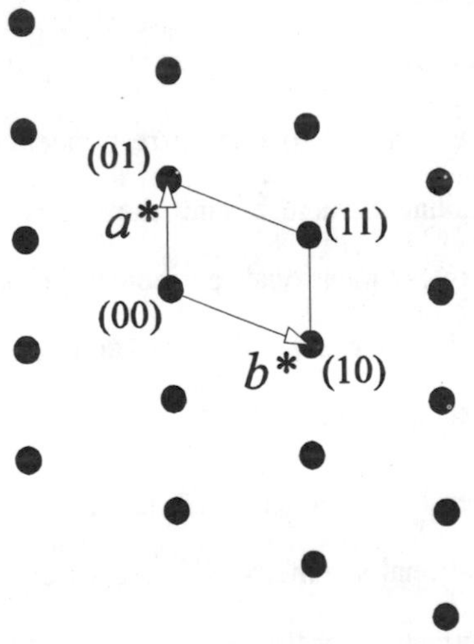

Fig. 1.12(b) The corresponding reciprocal-space lattice with the unit-mesh vectors (***a****,***b****) being constructed according to Eq. (1.10).

We can construct the reciprocal lattice by defining first the reciprocal unit vectors as

$$\boldsymbol{a}^* = 2\pi\left(\frac{\boldsymbol{b}\times\mathbf{n}}{\boldsymbol{a}\cdot(\boldsymbol{b}\times\mathbf{n})}\right), \tag{1.10a}$$

$$\boldsymbol{b}^* = 2\pi\left(\frac{\mathbf{n}\times\boldsymbol{a}}{\boldsymbol{b}\cdot(\mathbf{n}\times\boldsymbol{a})}\right). \tag{1.10b}$$

Any reciprocal lattice vector can be represented by a vector $\mathbf{G}_{nm}$,

$$\mathbf{G}_{nm} = n\,\boldsymbol{a}^* + m\,\boldsymbol{b}^* . \tag{1.11}$$

The relationships between the unit vectors in the real space and the reciprocal space are $\boldsymbol{a}\cdot\boldsymbol{a}^*=\boldsymbol{b}\cdot\boldsymbol{b}^*=2\pi$ and $\boldsymbol{a}\cdot\boldsymbol{b}^*=\boldsymbol{b}\cdot\boldsymbol{a}^*=0$. Corresponding to the real space lattice shown in Fig. 1.12(a), the reciprocal space lattice structure can thus be plotted in Fig. 1.12(b).

One can show rigorously that the reciprocal lattice structure constructed above matches exactly the diffraction pattern for the corresponding real space lattice. For the specific cases discussed in the previous section, we have a crystalline surface with $\boldsymbol{a} = a\mathbf{e}_x$, $\boldsymbol{b} = b\mathbf{e}_y$ and $\mathbf{n} = \mathbf{e}_z$. The reciprocal lattice structure can be constructed based on the derived unit vectors, $\boldsymbol{a}^* = (2\pi/a)\mathbf{e}_x$ and $\boldsymbol{b}^* = (2\pi/b)\mathbf{e}_y$. Using Eq. (1.11) we can rewrite Eq. (1.9) of the diffraction intensity distribution from an infinitely large perfect surface as

$$\mathrm{S}(\,\mathbf{k}\,) \sim \sum_{n,\,m} \delta(\,\mathbf{k}_{\|} - \mathbf{G}_{nm}\,)\,.$$

The intensity spots are located at the reciprocal lattice positions, $\mathbf{G}_{nm} = \frac{2n\pi}{a}\mathbf{e}_x - \frac{2m\pi}{b}\mathbf{e}_y$, which is consistent with the diffraction patterns obtained in Figs. 1.8(b) and 1.11.

§I.3 Diffraction Characteristics of Continuous Surfaces

In a continuous surface, the discrete atomic structure is ignored. Such a surface model has been frequently used for the study of liquid or amorphous surfaces.[1.12] This model can also be used to describe the crystalline surface provided that the diffraction wavelength is much longer than the atomic length scale, as in the case of visible light scattering.[1.13]

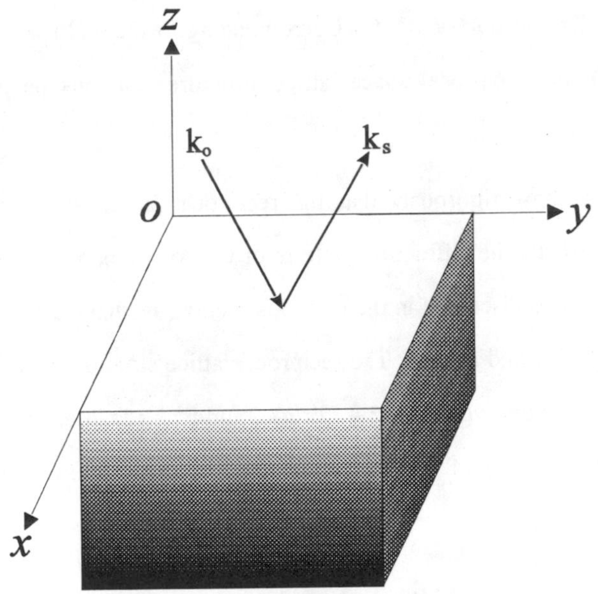

Fig. 1.13 Diffraction geometry from a continuous flat surface with a bulk structure.

Consider a continuous flat surface with dimensions, L_x, L_y and L_z, along the x, y and z directions, respectively, as shown in Fig. 1.13. The diffraction geometry is given in Fig. 1.13 where the incident beam is diffracted from the surface of the *xy* plane. For a continuous surface, the scattering density of the bulk is assumed to be a constant, e.g., $D(\rho) = 1$. Similar to the case in **§I.2.2**, if attenuation is considered, the diffraction amplitude is given by

$$A(\mathbf{k}) = \int e^{-\kappa|z|} e^{-i\mathbf{k}\cdot\rho} d^3\rho = \int_{-L_z}^{0} e^{\kappa z} e^{-ik_z z} dz \int_0^{L_y} e^{-ik_y y} dy \int_0^{L_x} e^{-ik_x x} dx.$$

We then obtain the diffraction intensity as

$$S(\mathbf{k}) = \Phi_c(k_z) \frac{\sin^2(L_x k_x/2)}{(k_x/2)^2} \times \frac{\sin^2(L_y k_y/2)}{(k_y/2)^2}. \quad (1.12)$$

Under the assumption, $L_z >> 1/\kappa$, we can write

$$\Phi_c(k_z) = \frac{1}{\kappa^2 + k_z^2}.$$

Again, the factor $\Phi_c(k_z)$ does not affect the relative intensity distribution measured by the detector in the k_x-k_y plane. For an extremely large system, Eq. (1.12) becomes

$$S(\mathbf{k}) \sim \Phi_c(k_z)\, \delta(\mathbf{k}_{\parallel}),$$

where we have used the identity,

$$\frac{\sin^2(LQ/2)}{(Q/2)^2} \sim \delta(Q), \quad \text{as } L \to \infty.$$

The reciprocal space characteristics of Eq. (1.12) are summarized in Fig. 1.14. Similar to the previous description, the size of the rod shown in Fig. 1.14 equals the FWHM of the intensity distribution and is inversely proportional to the surface size, L_x and L_y. However, in contrast to the case of the crystalline surface shown in Fig. 1.8(a), the continuous diffraction only has one rod located at the origin, (k_x=0, k_y=0). The continuous symmetry of the surface has eliminated all the higher order diffraction beam intensities that occur in the crystalline surface. The physical reason is that the constructive interference of diffraction occurs only at $\mathbf{k}_{\|} = 0$ for a continuous surface. However, for a crystalline surface, the constructive interference can occur at any reciprocal lattice positions, $\mathbf{k}_{\|} = \mathbf{G}_{nm}$, due to the existence of the real space periodicity.

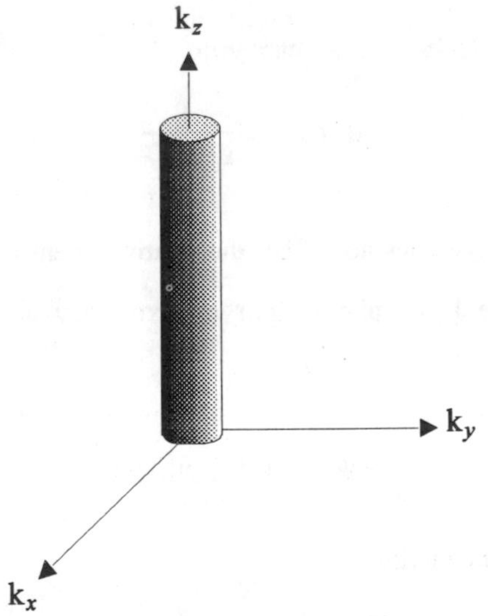

Fig. 1.14 Reciprocal-space characteristics of diffraction intensity distribution from a continuous surface.

§I.4 Diffraction Techniques

There are a number of diffraction techniques available for the study of surface roughness and dynamics of growth. The methods can be divided into three categories according to the nature of the quantum particles of the diffraction wave. The first type is electron diffraction, such as Low Energy Electron Diffraction (LEED) and Reflection High Energy Electron Diffraction (RHEED). The second type is electromagnetic radiation diffraction, including X-ray and diffuse light scattering. The third type is nuclear or atom diffraction, such as neutron, proton and atom scattering techniques.

For different techniques, the diffraction intensity expressions contain a common diffraction structure factor but have different prefactors. The diffraction structure factor is solely determined by

$$S(\mathbf{k}) \propto \left| \int D(\rho)\, e^{-i\mathbf{k}\cdot\rho}\, d^3\rho \right|^2 , \qquad (1.13)$$

which has been utilized frequently in the previous sections. However, the total diffraction intensity contains a prefactor which depends on the detailed interactions between the diffraction wave and the surface atoms. In electron diffraction, for example, the intensity can be written as

$$I(\mathbf{k}) = I_0(\mathbf{k})\, S(\mathbf{k}) ,$$

where the prefactor $I_0(\mathbf{k})$ is the atomic form factor depending on the electron-atom scattering cross section. For X-ray and neutron diffraction techniques, the Born approximation leads to a diffraction intensity expression, [1.12]

$$I(\mathbf{k}) \propto \left(\frac{b}{k_{\perp}}\right)^2 S(\mathbf{k}),$$

where for neutron scattering, b is the scattering length of the nuclei and for X-ray scattering, $b=(e^2/m^2)$, where e and m are the electron charge and rest mass, respectively. Usually, the prefactor is a slowly varying function of **k** as compared to the structure factor, S(**k**). Therefore, for our purpose, we will focus only on the surface diffraction structure factor, Eq. (1.13). The contribution of the prefactor in any specific diffraction technique will not be concerned in this book. As far as the shape is concerned, S(**k**) and I(**k**) are considered to be identical.

The major diffraction techniques relevant to the study of the rough surface are shown in table (I-1.). In the table we also make some comments on the advantages and disadvantages of each technique and its major areas of applications in recent years.

Instrumental resolution is a very important factor in determining the performance and capability of the techniques. The narrower the instrument response width the better the resolution. It limits the largest area on the surface that the technique can be used to gain the structural information of the surface. The rigorous definition of the resolving power will be discussed in more detail in the next section. All the techniques can provide a real space resolution larger than 1000 Å except atom diffraction. RHEED has a directional resolution in the sense that it is very high in the scattering plane but poor in the plane perpendicular to the scattering plane.

For dynamic growth study, temporal resolution is also an important factor. In this regard both electron diffraction and the light scattering can provide sufficient counts in a short time. In the case of HRLEED, for example, one can obtain as high as 10^4 counts in a millisecond at a particular angular position. But both X-ray diffraction (including the intense X-ray beam generated from the Synchrotron radiation) and atom diffraction would have difficulty in providing sufficient intensity in this short time scale.

Table I-1. Comparison among different diffraction techniques

Techniques	Advantages	Disadvantages	Applications
HRLEED	High spatial resolution ($\sim 10^{-2}$–10^{-3} Å^{-1}) High temporal resolution ($\sim 10^{-3}$s) Very large k_z range	Multiple scattering(?)	Rough crystalline surface Epitaxial growth dynamics
RHEED	High spatial resolution ($\sim 10^{-3}$ Å^{-1}) High temporal resolution Very large k_z range	Multiple scattering(?)	Rough crystalline surface Epitaxial growth dynamics
X-ray	High spatial resolution ($\sim 10^{-3}$ Å^{-1}) Single scattering	Low counting rate for temporal resolution Small k_z range	Rough crystalline and non-crystalline surfaces Multilayer thin films
Atom diffraction	Single scattering Reasonable resolution ($> 10^{-2}$ Å^{-1}) Reasonable k_z range	Large thermal diffuse scattering Poor off-(00) beams	Rough crystalline surface
Diffuse light scattering	High resolution ($\sim 10^{-3}$ Å^{-1})	Small k_z range	Rough crystalline and non-crystalline surfaces

In addition to the intensity (Counts/Sec.) of the diffraction beam, the sweeping speed for the measurement of the angular profile is a very important consideration for high temporal resolution. In this regards, position sensitive detectors, such as a 2D electron channel plate, video LEED system, and position sensitive photon (light or X-ray) detectors, would be very powerful and can reduce significantly the data acquisition time for the angular profile measurement. The conventional scan using mechanical means would not be desirable for growth studies in short time scales. For electron diffraction, electrostatic or magnetic scan of the beam (fixed detector) can also eliminate the delay going from one angular position to the next in the angular scan of the profile.

There are three parameters associated with a rough surface that one would like to measure: the roughness parameter α, the vertical correlation length, and the lateral correlation length. (These terms will be defined later.) As we shall see later, in order to measure all three parameters, one requires intensity measurements over a wide range of k_Z. Thus far, electron diffraction (HRLEED or HREED) can provide the largest k_Z range for the surface roughness study. Other diffraction techniques either have difficulty or can not reach a wide range of k_Z.

On the other hands, electrons interact very strongly with the surface so that the intensity contains a large number of multiple scattering events which may obscure the simple interpretation of the results. Other techniques such as X-ray diffraction would have no such problem. However, one very important fact about electron diffraction is that, at least in the HRLEED geometry, the effect of multiple scattering on the shape of the angular distribution of the diffraction intensity (proportional to the diffraction structure factor) is not significant enough to be detectable. A remarkable recent example [1.14, 1.15] is the HRLEED measurements of the surface roughening transition in which the shape of the diffraction structure factor changed dramatically at the "out-of-phase" diffraction condition (to be defined later) on going through the transition due to the generation of a high density of surface steps. During this transition the structure factor stayed the same at the "in-phase" diffraction condition. We conclude that multiple scattering must have negligible effect on the shape of the structure factor and the change of the structure factor at the "out-of-phase" condition must solely come from the effect of generation of steps. Fortunately, all the important information on the rough surface is obtained either through the shape of the diffraction structure factor, or the relative change of the intensity, but NOT the absolute intensity. Electron diffraction therefore remains an extremely important tool for the study of rough surfaces and dynamics of growth.

§I.5 Instrument Response

Even for a perfect large surface in which the structure factor is a δ–function, one would never obtain such a function in an actual measurement due to the finite resolution of the instrument. The measured diffraction intensity profile would be broadened due to instrumental effects such as finite beam size, beam divergence, energy spread, and detector width. [1.16, 1.17] The net result is that even if we have a perfect surface one would give a broadened diffraction profile with a finite width called the instrument response. One can define the instrument response as a function, $T(\mathbf{k},\mathbf{k}')$, by which the experimentally measured diffraction intensity, $I_e(\mathbf{k})$, is the convolution of the true intensity and the instrument response:

$$I_e(\mathbf{k}) = T(\mathbf{k}, \mathbf{k}') * I(\mathbf{k}') = \int T(\mathbf{k}, \mathbf{k}')\, I(\mathbf{k}')\, d^2k',$$

where **k** is the experimentally measured momentum transfer.

Very often the instrument response can be well represented by a Gaussian function,

$$T(\mathbf{k}, \mathbf{k}') \propto \exp\left(-\frac{(\mathbf{k}-\mathbf{k}')^2}{\sigma_G^2}\right),$$

where the width of the function, σ_G, depends in general on the incident beam energy and the diffraction geometry.

If one can measure the shape of the functions $I_e(\mathbf{k})$ and $T(\mathbf{k}, \mathbf{k}')$ precisely, then in principle the true intensity (therefore the diffraction structure factor) can be obtained precisely either by a deconvolution process or by iteratively convoluting trial functions $I(\mathbf{k}')$ with $T(\mathbf{k}, \mathbf{k}')$ to fit $I_e(\mathbf{k})$. A simple example is when both $I_e(\mathbf{k})$ and $T(\mathbf{k}, \mathbf{k}')$ are Gaussians, i.e.,

$$I_e(\mathbf{k}) \propto \exp\left(-\frac{\mathbf{k}^2}{\sigma_e^2}\right),$$

where σ_e is the measured width of the diffraction intensity distribution, $I_e(\mathbf{k})$. Then the true intensity $I(\mathbf{k})$ must be a Gaussian function, $\exp\left(-\frac{\mathbf{k}^2}{\sigma^2}\right)$, because

$$I_e(\mathbf{k}) \propto \int \exp\left(-\frac{(\mathbf{k}-\mathbf{k}')^2}{\sigma_G^2}\right) \exp\left(-\frac{\mathbf{k}'^2}{\sigma^2}\right) d^2k' \propto \exp\left(-\frac{\mathbf{k}^2}{\sigma^2+\sigma_G^2}\right) = \exp\left(-\frac{\mathbf{k}^2}{\sigma_e^2}\right).$$

That is, $\sigma^2 + \sigma_G^2 = \sigma_e^2$, or $\sigma^2 = \sigma_e^2 - \sigma_G^2$. Thus, $I(\mathbf{k})$ can be determined precisely.

However, in practice, one can never measure $I_e(\mathbf{k})$ and $T(\mathbf{k},\mathbf{k}')$ with infinite precision. As a result, an uncertainty will be introduced when determining the $I(\mathbf{k})$

function. Even for the case of Gaussian profiles, an uncertainty will be introduced when determining the width of T(**k**,**k**') and I_e(**k**). An uncertainty therefore always exists in the determination of I(**k**). The uncertainty in the determination of I(**k**) can be reduced by[1.2, 1.18]: (1) measuring the same profile many times and taking the average, thus reducing the percentage error in the measurement; and (2) making a better instrument to reduce the width of T(**k**, **k**'). Very often, researchers use just the width of T(**k**, **k**') as an indication of the resolving power of the instrument and use the inverse of T(**k**, **k**'), that is,

$$\frac{2\pi}{\text{width of } T(\mathbf{k}, \mathbf{k}')},$$

as the largest surface area in real space (coherent width) from which one can extract surface structural information. This scheme can serve as a crude estimate of the resolving capability of the instrument. One can do much better if one takes into account the first strategy stated in (1).[1.2, 1.18]

REVIEW AND SUMMARY

Diffraction structure factor

Diffraction intensity is proportional to a common diffraction structure factor given by

$$S(\mathbf{k}) = \left| \int D(\rho)\, e^{-i\mathbf{k}\cdot\rho}\, d^3\rho \right|^2 .$$

D(ρ) is the density function of scatterers in a medium. **k** is the momentum transfer defined as the difference between the incident and the out-going wave vectors, $\mathbf{k} = \mathbf{k}_S - \mathbf{k}_0$.

2D reciprocal lattice

For a 2D real-space lattice with a normal direction **n** and unit-mesh vectors given by (***a***, ***b***), one can construct a 2D reciprocal-space lattice with the unit-mesh vectors (***a****, ***b****) defined by

$$\boldsymbol{a}^* = 2\pi \left(\frac{\boldsymbol{b} \times \mathbf{n}}{\boldsymbol{a}\cdot(\boldsymbol{b} \times \mathbf{n})} \right), \quad \boldsymbol{b}^* = 2\pi \left(\frac{\mathbf{n} \times \boldsymbol{a}}{\boldsymbol{b}\cdot(\mathbf{n} \times \boldsymbol{a})} \right). \qquad (1.10)$$

The reciprocal-space lattice vectors,

$$\mathbf{G}_{nm} = n\, \boldsymbol{a}^* + m\, \boldsymbol{b}^* , \qquad (1.11)$$

determine the LEED patterns in diffraction experiments.

Diffraction from a perfectly flat crystalline surface

The diffraction structure factor is a sum of sharp δ–functions located at the reciprocal lattice positions,

$$S(\mathbf{k}) \sim \sum_{n,\, m} \delta(\mathbf{k}_{||} - \mathbf{G}_{nm}) ,$$

where $\mathbf{k}_{||}$ is the momentum transfer parallel to the surface.

Diffraction from a perfectly flat continuous surface

The diffraction structure factor contains one sharp δ–function located at $\mathbf{k}_{\|} = 0$,

$$S(\mathbf{k}) \sim \delta(\mathbf{k}_{\|}) .$$

REFERENCES

1.1 C. Kittel, *Introduction to Solid State Physics* (John Wiley and Sons, Inc, New York, 1986), Chapter 2.

1.2 M. G. Lagally in *Methods of Experimental Physics*, Vol. **22** (Academic Press, New York, 1985), pp. 237; M. G. Lagally and J. A. Martin, *Rev. Sci. Instrum*. **54**, 1237 (1983).

1.3 M. A. Van Hove, W. H. Weinberg, and C.-M. Chan, "Low-energy electron diffraction", *Springer Ser. Surf. Sci.*, Vol. **6** (Springer-Verlag, Berlin, 1986), pp. 137.

1.4 U. Scheithauer, G. Meyer, and M. Henzler, *Surf. Sci.* **178**, 441 (1986).

1.5 J.-K. Zuo, R. A. Harper, and G.-C. Wang, *Appl. Phys. Lett.* **51**, 250 (1987).

1.6 O. Stern, Naturwissenschaften **17**, 391 (1929).

1.7 For a review, see T. Engel and K. H. Rieder in "Structural studies of surfaces", *Springer Tracts in Modern Physics*, Vol. **31** (Springer-Verlag, Berlin, 1982), pp. 55.

1.8 For a recent review, see I. K. Robinson, *Reports on Progress in Physics* **55**, 599 (1992).

1.9 J. Als-Nielsen in "Structure and dynamics of surfaces II", *Topics in Current Physics* Vol. **43** (Springer-Verlag, Berlin, 1987), pp. 181.

1.10 R. Feidenhans'l, *Surf. Sci. Rep.* **10**, 105 (1989).

1.11 For a recent review, see "Reflection high-energy electron diffraction and reflection electron imaging of surfaces", ed. P. K. Larsen and P. J. Bobson, *NATO ASI Series B: Physics,* Vol. **188** (Plenum Press, New York, 1988), pp. 3.

1.12 S. K. Sinha, E. B. Sirota, S. Garoff, and H. B. Stanley, *Phys. Rev.* **B38**, 2297 (1987).

1.13 For a review, see J. M. Bennett and L. Mattsson, *Introduction to Surface Roughness and Scattering* (Optical Society of America, Washington, D.C., 1989).

1.14 H.-N. Yang, T.-M. Lu, and G.-C. Wang, *Phys. Rev. Lett.* **63**, 1621 (1989).

1.15 H.-N. Yang, T.-M. Lu, and G.-C. Wang, *Phy. Rev.* **B43**, 4714 (1991).

1.16 R. L. Park, J. E. Houston, and D. G. Schreiner, *Rev. Sci. Instrum.* **42**, 60 (1971).

1.17 G.-C. Wang and M. G. Lagally, *Surf. Sci.* **81**, 69 (1979).

1.18 T.-M. Lu and M. G. Lagally, *Surf. Sci.* **99**, 695 (1980).

Chapter II CORRELATION FUNCTIONS FOR ROUGH SURFACES

In this Chapter we shall define the diffraction structure factor, which is proportional to the angular distribution of the diffraction intensity, from a rough surface with height variation as a function of the position on the surface. It is shown that the diffraction structure factor can be written as the Fourier transform of the height difference function which describes the statistical average of the phase difference caused by the height variation of the surface. Both crystalline and non-crystalline surfaces are considered. Based on the Gaussian approximation, the height difference function can be written in terms of a height-height correlation function which describes the characteristics of the rough surface. The height-height correlation function contains three parameters: lateral correlation length, vertical correlation length, and the roughness parameter α. Some general properties of the height-height correlation function are discussed.

§II.1 Diffraction Structure Factor

In the last chapter (§I.4), we introduce the concept of the diffraction structure factor, $S(\mathbf{k})$, given by Eq. (1.13). We also show results of the calculation for simple examples (see §I.2.2 and §I.3). In those examples, the surface density function $D(\boldsymbol{\rho})$ depends only on the 2-D positional vector, $\boldsymbol{\rho}$. This is because the surfaces are considered to be "flat". In this section, we consider a more general definition of the diffraction structure factor for rough surfaces with corrugations. The density function $D(\boldsymbol{\rho}, z)$ depends not only on $\boldsymbol{\rho}$, but also on the vertical position z.

§II.1.1 *Diffraction structure factor from a continuous surface*

Consider a small section of a continuous surface, as shown in Fig. 2.1. The normal direction for the average surface plane is always assumed to be along the z axis. For a continuous surface, the surface height variation can be described by a height function, $z = z(\rho)$. The surface position is denoted by the lateral coordinate $\rho = (x,y)$ and vertical coordinate z. Throughout this monograph, we shall always assume that the lateral size of a surface is sufficiently large so that the finite size effect or boundary conditions will not be a concern. This assumption is valid for most realistic systems.

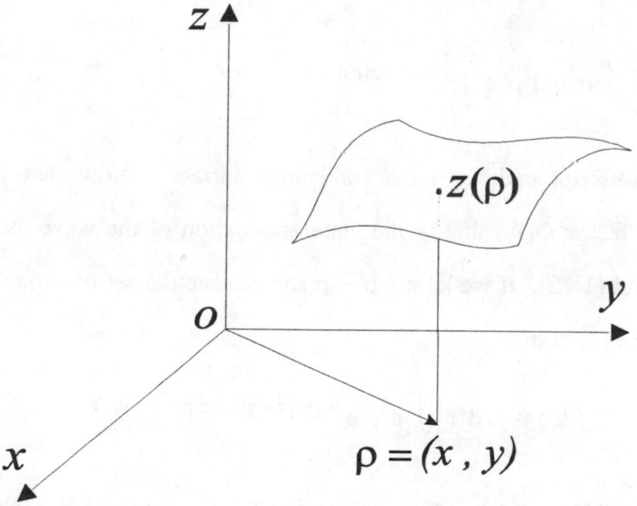

Fig. 2.1 The geometry of a continuous surface in a Cartesian coordinate system. The normal direction for the average surface plane is assumed to be along the z axis.

The surface density function for a continuous surface is expressed as

$$D(\rho, z') = \delta[z' - z(\rho)] \; .$$

The corresponding diffraction amplitude is given by

$$A(\mathbf{k}) = \int D(\rho, z')\, e^{-i k_\perp z'}\, e^{-i \mathbf{k}_\parallel \cdot \rho}\, dz'\, d^2\rho$$

$$= \int d^2\rho\, e^{-i k_\perp z(\rho)}\, e^{-i \mathbf{k}_\parallel \cdot \rho},$$

where $k_\perp \equiv k_z$ is the component of the wavevector $\mathbf{k}$ perpendicular to the *x-y* plane. Accordingly, we obtain the diffraction structure factor as

$$S_c(\mathbf{k}) = A^*(\mathbf{k})\, A(\mathbf{k}) = \int d^2\rho'\, e^{ik_\perp z(\rho')}\, e^{i\mathbf{k}_\parallel \cdot \rho'} \int d^2\rho\, e^{-ik_\perp z(\rho)}\, e^{-i\mathbf{k}_\parallel \cdot \rho}$$

$$= \int \int d^2\rho\, d^2\rho'\, e^{ik_\perp [z(\rho') - z(\rho)]}\, e^{i\mathbf{k}_\parallel \cdot (\rho' - \rho)}, \qquad (2.1)$$

where the subscript c denotes the continuous surface. Note that we have not included the factor $\Phi(k_z)$ due to the finite attenuation of the wave as described in Eqs. (1.7') and (1.12). If we let $\mathbf{r} = \rho' - \rho$ and replace the set of variables (ρ, ρ') by $(\mathbf{r}, \rho)$, Eq. (2.1) becomes

$$S_c(\mathbf{k}) = \int d^2r \left(\int d^2\rho\, e^{ik_\perp [z(\mathbf{r} + \rho) - z(\rho)]} \right) e^{i\mathbf{k}_\parallel \cdot \mathbf{r}}. \qquad (2.2)$$

Since the lateral size of the surface is extremely large, we have ignored the boundary effect on the integral due to the transformation of variables. The integration in the bracket of Eq. (2.2) is only a function of $k_\perp$ and the lateral distance, $\mathbf{r}$. Such an integral over a sufficiently large system will give a statistical average quantity called the height difference correlation function, $C_c(k_\perp, \mathbf{r})$, i.e.,

$$\frac{1}{A}\int d^2\rho \; e^{\,ik_\perp[z(\mathbf{r}+\rho)-z(\rho)]} = \langle e^{\,ik_\perp[z(\mathbf{r})-z(0)]} \rangle = C_c(k_\perp, \mathbf{r}), \qquad (2.3)$$

where A is the lateral area of the surface, $A = \int d^2\rho$. Combining Eq. (2.2) with Eq. (2.3), we can define rigorously the diffraction structure factor from a continuous surface as

$$S_c(\mathbf{k}) = \int d^2r \; C_c(k_\perp, \mathbf{r}) \, e^{\,i\mathbf{k}_\parallel \cdot \mathbf{r}} = \int d^2r \langle e^{\,ik_\perp[z(\mathbf{r})-z(0)]} \rangle \, e^{\,i\mathbf{k}_\parallel \cdot \mathbf{r}}, \qquad (2.4)$$

which is proportional to the intensity distribution. (We ignore the constant A.)

We can now see from Eq. (2.4) that the diffraction structure factor actually is a Fourier transform of the height difference function defined by Eq. (2.3) which is a measure of the statistical phase correlation between surface positions. The measurement is distinctly different from that of real-space imaging techniques, such as scanning electron microscopy (SEM) or scanning tunneling microscopy (STM), in which one measures directly the surface topography, $z = z(\rho)$. Thus, in diffraction techniques one measures the statistical average properties of the rough surface.

§II.1.2 *Diffraction structure factor from a crystalline surface*

Similar to the case of the continuous surface shown in Fig. 2.1, the height variation in a crystalline surface can be described by a function, $z = z(\rho) = h(\rho)c$, where $h(\rho)$ is a measure of the crystalline surface height in units of the single step height (or vertical spacing), c. The function $h(\rho)$ must take an integer value at the lattice point, $\rho_{mn} = m\mathbf{a} + n\mathbf{b}$. The density function for this periodic crystalline surface is expressed as

$$D(\rho, z) = \sum_{m,n} \delta(\rho - \rho_{mn})\, \delta[z - h(\rho)c]$$

$$= F(\rho)\, \delta[z - h(\rho)c],$$

where F(ρ) represents the atomic density of a perfect crystalline surface (lateral) and can be called the two-dimensional lattice function,

$$F(\rho) \equiv \sum_{m,n} \delta(\rho - \rho_{mn}). \qquad (2.5)$$

With the calculation of the diffraction amplitude,

$$A(\mathbf{k}) = \int D(\rho, z)\, e^{-i k_{\perp} z}\, e^{-i \mathbf{k}_{\parallel} \cdot \rho}\, dz\, d^2\rho$$

$$= \int d^2\rho\, e^{-i k_{\perp} c\, h(\rho)}\, e^{-i \mathbf{k}_{\parallel} \cdot \rho}\, F(\rho),$$

we are able to follow the same procedures as that in the continuous surface case discussed in §II.1.1 to derive the diffraction structure factor, i.e.,

$$S_d(\mathbf{k}) = A^*(\mathbf{k}) A(\mathbf{k})$$

$$= \int d^2\rho'\, e^{i k_{\perp} c\, h(\rho')}\, e^{i \mathbf{k}_{\parallel} \cdot \rho'}\, F(\rho') \int d^2\rho\, e^{-i k_{\perp} c\, h(\rho)}\, e^{-i \mathbf{k}_{\parallel} \cdot \rho}\, F(\rho)$$

$$= \int \int d^2\rho\, d^2\rho'\, e^{i k_{\perp} c\, [h(\rho') - h(\rho)]}\, F(\rho) F(\rho')\, e^{i \mathbf{k}_{\parallel} \cdot (\rho' - \rho)}, \qquad (2.6)$$

where the subscript, d, denotes the discrete nature of a crystalline surface. Let **r** = ρ' − ρ and replace the set of variables (ρ, ρ') by (**r**, ρ). We find

$$S_d(\mathbf{k}) = \int d^2r \left(\int d^2\rho \, e^{ik_\perp c\,[h(\mathbf{r}+\rho) - h(\rho)]} F(\rho) F(\mathbf{r}+\rho) \right) e^{i\mathbf{k}_\parallel \cdot \mathbf{r}}. \qquad (2.7)$$

The integral in the bracket can be further calculated using Eq. (2.5),

$$\int d^2\rho \, e^{ik_\perp c\,[h(\mathbf{r}+\rho) - h(\rho)]} F(\rho) F(\mathbf{r}+\rho) =$$

$$= \int d^2\rho \, e^{ik_\perp c\,[h(\mathbf{r}+\rho) - h(\rho)]} \sum_{m,n} \delta(\rho - \rho_{mn}) F(\mathbf{r}+\rho)$$

$$= \sum_{m,n} e^{ik_\perp c\,[h(\rho_{mn}+\mathbf{r}) - h(\rho_{mn})]} F(\mathbf{r}+\rho_{mn})$$

$$= \left(\sum_{m,n} e^{ik_\perp c\,[h(\mathbf{r}+\rho_{mn}) - h(\rho_{mn})]} \right) F(\mathbf{r}), \qquad (2.8)$$

where we have used the identity, $F(\mathbf{r} + \rho_{mn}) = F(\mathbf{r})$, based on the periodicity of the two-dimensional lattice shown in Eq. (2.5). Compared with Eq. (2.3) for the continuous surface case, the term in brackets in Eq. (2.8) is directly related to the discrete version of the height difference function, i.e.,

$$\frac{1}{N} \sum_{m,n} e^{ik_\perp c\,[h(\mathbf{r}+\rho_{mn}) - h(\rho_{mn})]} = \langle e^{ik_\perp c\,[h(\mathbf{r}) - h(0)]} \rangle = C_d(k_\perp, \mathbf{r}), \qquad (2.9)$$

where N is the total number of the scatterers in the surface. Combining Eq. (2.7) with Eqs. (2.8) and (2.9), we can write the diffraction structure factor, Eq. (2.7), as

$$S_d(\mathbf{k}) = N \int d^2r \, C_d(k_\perp, \mathbf{r}) \, F(\mathbf{r}) \, e^{i\mathbf{k}_\parallel \cdot \mathbf{r}}. \qquad (2.10)$$

Compared with Eq. (2.4), the integration in Eq. (2.10) contains a convolution function, F(**r**), which results from the discrete nature of a crystalline surface. If we insert Eq. (2.5) into Eq. (2.10) and ignore the constant N, we can define the discrete version of the diffraction structure factor as

$$S_d(\mathbf{k}) = \int d^2r\, C_d(k_\perp, \mathbf{r})\, F(\mathbf{r})\, e^{i\mathbf{k}_\parallel \cdot \mathbf{r}}$$

$$= \sum_{m,n} C_d(k_\perp, \rho_{mn})\, e^{i\mathbf{k}_\parallel \cdot \rho_{mn}}$$

$$= \sum_{m,n} < e^{ik_\perp c\,[h(\rho_{mn}) - h(0)]} > e^{i\mathbf{k}_\parallel \cdot \rho_{mn}}\,. \qquad (2.11)$$

By analogy to the continuous surface case, $S_d(\mathbf{k})$ represents the discrete Fourier transform of the height difference correlation function, Eq. (2.9).

Equation (2.11) can have a different expression, if we employ an identity,

$$F(\rho) = \sum_{m,n} \delta(\rho - \rho_{mn}) = \frac{1}{v}\sum_{k,l} e^{-i\mathbf{G}_{kl}\cdot\rho}\,, \qquad (2.12)$$

where $v = |\mathbf{a} \times \mathbf{b}|$ denotes the area of the real-space lattice unit mesh and $G_{kl} = k\,\mathbf{a}^* + l\,\mathbf{b}^*$ is the two-dimensional reciprocal lattice vector defined in §I.2.3. The proof of Eq. (2.12) is given in Appendix (IIA). Inserting Eq. (2.12) into Eq. (2.10), we rewrite the discrete diffraction structure factor as

$$S_d(\mathbf{k}) = \int d^2r\, C_d(k_\perp, \mathbf{r})\, F(\mathbf{r})\, e^{i\mathbf{k}_\parallel \cdot \mathbf{r}} = \frac{1}{v}\sum_{k,l} \int d^2r\, C_d(k_\perp, \mathbf{r})\, e^{i(\mathbf{k}_\parallel - \mathbf{G}_{kl})\cdot \mathbf{r}}. \qquad (2.13)$$

The integral in Eq. (2.13),

$$S'_d(\mathbf{k}) = \int d^2r\, C_d(k_\perp, \mathbf{r})\, e^{i\mathbf{k}_\parallel \cdot \mathbf{r}}, \qquad (2.14)$$

has a form similar to $S_c(\mathbf{k})$ for a continuous surface shown in Eq. (2.4). Equation (2.13) thus becomes

$$S_d(\mathbf{k}) = \frac{1}{V}\sum_{k,l} S'_d(\mathbf{k} - G_{kl}). \qquad (2.15)$$

We must emphasize that $S'_d(\mathbf{k})$ is distinctly different from $S_c(\mathbf{k})$ even though they have a similar expression. The discrete nature of a crystalline surface has significant effects on the diffraction structure factor in both the vertical and the lateral directions in reciprocal space. The "vertical" discrete effect manifests itself in the height difference functions, $C_d(k_\perp, \mathbf{r})$ shown as Eq. (2.9), and in $S'_d(\mathbf{k})$, Eq. (2.14). As will be shown later, $S'_d(\mathbf{k})$ is a periodic function of $k_\perp$ while the continuous surface structure factor, $S_c(\mathbf{k})$, is not. On the other hand, the "lateral" discrete effect from a crystalline surface manifests itself in the summation of $S'_d(\mathbf{k})$ in Eq. (2.15). The diffraction structure factor of a continuous surface shown in Eq. (2.4) has a line shape, $S_c(\mathbf{k})$, centered at $\mathbf{k}_\parallel = 0$. In contrast, the diffraction structure factor $S_d(\mathbf{k})$ in a crystalline surface is the superimposition of the components, $S'_d(\mathbf{k} - G_{kl})$, centered at different reciprocal vector positions, G_{kl}, corresponding to different Bragg diffraction positions. $S_d(\mathbf{k})$ is thus a periodic function of both $k_\perp$ and $\mathbf{k}_\parallel$.

For realistic application, we can simplify Eq. (2.15) considerably. Usually, the line shape distribution, $S'_d(\mathbf{k} - G_{kl})$, is confined to the region close to the Bragg beam positions, for example, within 10% of the Brillouin zone, for most experiments. This means that at the vicinity of the (*kl*) beam, we can approximate Eq. (2.15) by

$$S_d(\mathbf{k}) \approx \frac{1}{V} S'_d(\mathbf{k} - G_{kl}). \qquad (2.15')$$

The approximation can significantly simplify the calculation of $S_d(\mathbf{k})$ because the integration in Eq. (2.14) sometimes can give an analytical form but the summation in Eq. (2.11) usually does not.

In conclusion, the diffraction structure factor for both crystalline surfaces and continuous surfaces can be expressed as

$$S(\mathbf{k}) = \int d^2r\, C(k_\perp, \mathbf{r})\, e^{i\mathbf{k}_\parallel \cdot \mathbf{r}}, \qquad (2.16)$$

at the vicinity of the *(kl)* diffraction beam position. The corresponding height difference function $C(k_\perp, \mathbf{r})$ is given by Eq. (2.3) (for a continuous surface) or Eq. (2.9) (for a crystalline surface).

§II.2 Height Difference Function

As we have pointed out in the last section, the diffraction structure factor is solely determined by the height difference function, $C(k_\perp,\mathbf{r})$, which describes the statistical average of the phase difference caused by the height variation of the surface. In order to calculate $C(k_\perp,\mathbf{r})$, one needs to know the statistical distribution function, $g(\Delta z, \mathbf{r})$, for the relative surface height, $\Delta z = z(\mathbf{r}) - z(0)$, where $g(\Delta z, \mathbf{r})$ depends on the surface morphology. The height difference function, $C(k_\perp,\mathbf{r})$, is thus the Fourier transform of the distribution function, $g(\Delta z, \mathbf{r})$, with respect to $\Delta z = z(\mathbf{r}) - z(0)$,

$$C(k_\perp, \mathbf{r}) = \langle e^{ik_\perp[z(\mathbf{r}) - z(0)]} \rangle = \int_{-\infty}^{+\infty} d(\Delta z)\, g(\Delta z, \mathbf{r}) e^{ik_\perp \Delta z}. \qquad (2.17)$$

The height difference function $C(k_\perp, \mathbf{r})$ is also called the characteristic function of $g(\Delta z,\mathbf{r})$.

For a statistically rough surface, $g(\Delta z, \mathbf{r})$ can very often be represented by a Gaussian function in the following form,

$$g(\Delta z, \mathbf{r}) \sim e^{-\frac{(\Delta z)^2}{2\sigma^2}},$$

where the Gaussian width σ is a function of $\mathbf{r}$. In Appendix VIA, we show analytically how such distribution arises in a model surface containing a one-dimensional Markovian distribution of steps. [2.1] This model has been used quite extensively in the literature.

A Gaussian distribution is a valid approximation for many realistic rough surfaces. To give an example, we analyze the experimental data from the rough surface of an Au:Pd thin film deposited on a mica surface. [2.2] Figure 2.2(a) shows the surface topography of the Au:Pd film deposited on a mica surface measured by Scanning Tunneling Microscopy (STM) in a scale of 2400 Å × 2400 Å. Figure 2.2(b) is a one-dimensional STM scan corresponding to a cross section of Fig. 2.2(a). This rough surface image exhibits a typical mountain-valley-like morphology with island (or mountain) size ($\sim \xi$) and the vertical amplitude ($\sim w$) for the surface height fluctuation. ξ and w are also called the lateral correlation length and the vertical correlation length (or the interface width), respectively. These terms will be explained in more detail later. We plot the measured relative height distribution function, $g(\Delta z, \mathbf{r})$, at three distances, $r = 16.4$ Å, 28.1 Å and 60.8 Å, in Figs. 2.3(a), 2.3(b) and 2.3(c), respectively. All the measured data points (open squares) can be fitted well by Gaussian distribution functions (the solid curves).

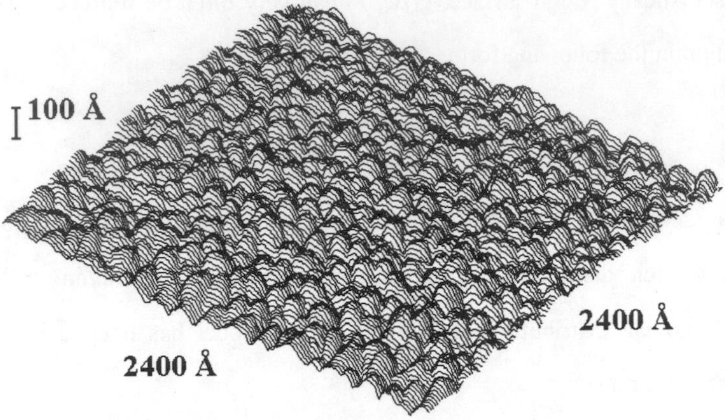

Fig. 2.2(a) The STM topography of a Au:Pd film deposited on a mica surface.[2.2]

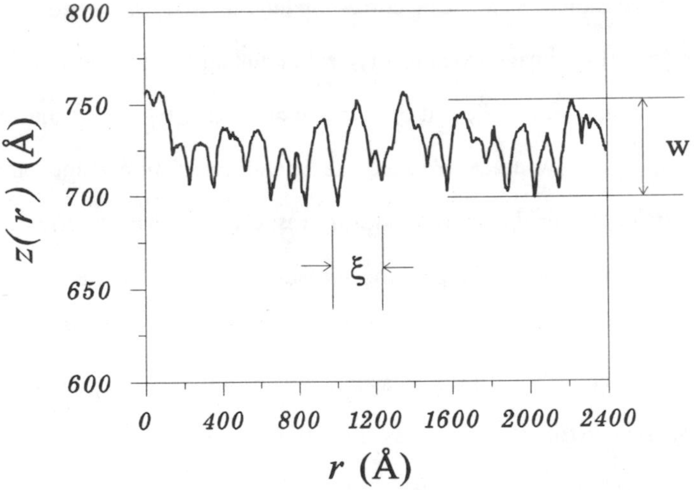

Fig. 2.2(b) A one-dimensional cross section of the STM topography shown in Fig. 2.2(a).

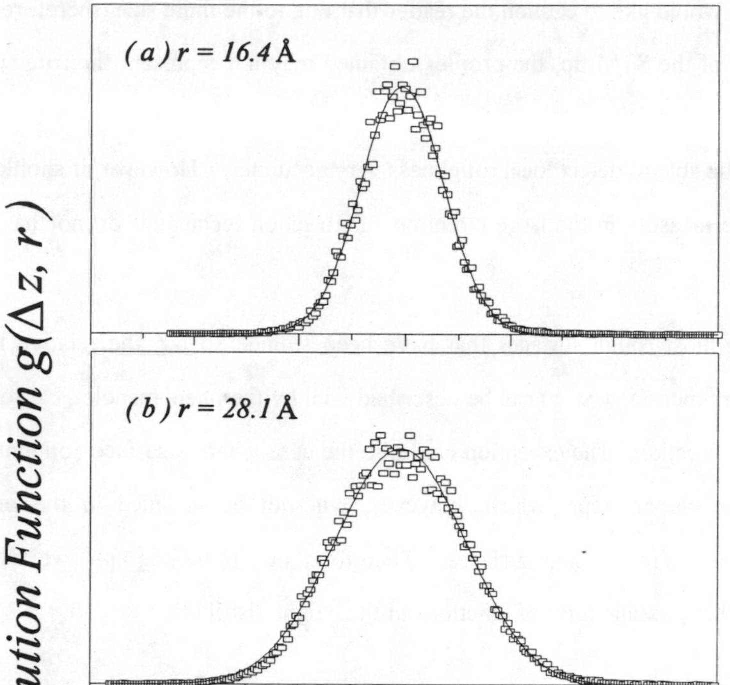

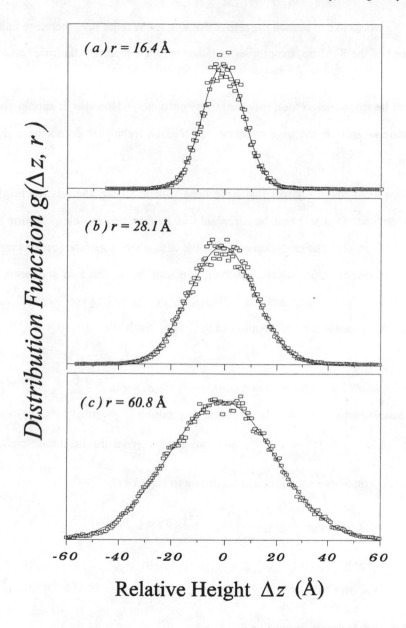

Fig. 2.3 Relative height distribution $g(\Delta z, \mathbf{r})$ at (a) r = 16.4 Å; (b) r = 28.1 Å; (c) r = 60.8 Å. The solid curves are the corresponding Gaussian functions.

We would like to caution the reader that due to the finite size (therefore finite resolution) of the STM tip, the profiles obtained may not represent the true surface morphology, especially in the small r regime. [2.3 - 2.5] The tip is not infinitely sharp and it may not be able to detect local roughness very accurately. However, it should give an accurate measure in the large r regime. Diffraction techniques do not have this shortcoming.

For most rough surfaces that have been studied so far, the relative height distribution function g(Δz, **r**) can be described well by the phenomenological form of a Gaussian function. The exception could be the case where a surface contains only very few levels of steps, which, however, will not be included in the present treatment of multilevel step surfaces. Therefore, in this monograph, we will not consider other possible form of functions in the height distribution.

§II.2.1 *Height difference function of a continuous rough surface*

Under the assumption that the relative height distribution is governed by a Gaussian function, $g(\Delta z, \mathbf{r}) = \frac{1}{\sigma\sqrt{2\pi}} e^{-\frac{(\Delta z)^2}{2\sigma^2}}$, one can easily derive the height difference function of a continuous rough surface according to Eq. (2.17),

$$C_c(k_\perp, \mathbf{r}) = \int_{-\infty}^{+\infty} d(\Delta z) \frac{1}{\sigma\sqrt{2\pi}} e^{-\frac{(\Delta z)^2}{2\sigma^2}} e^{ik_\perp \Delta z} = e^{-\frac{1}{2}(k_\perp \sigma)^2} .$$

The standard deviation, σ, determined by $\sigma^2 = \int_{-\infty}^{+\infty} (\Delta z)^2 \frac{1}{\sigma\sqrt{2\pi}} e^{-\frac{(\Delta z)^2}{2\sigma^2}} d(\Delta z) = \langle(\Delta z)^2\rangle$, is called the height-height correlation function,

$$H_c(\mathbf{r}) = \sigma^2 = \langle (\Delta z)^2 \rangle = \langle [z(\mathbf{r}) - z(0)]^2 \rangle . \qquad (2.18)$$

The height-height correlation function is a measure of the relative height fluctuation in a rough surface. The height difference function in a continuous rough surface can thus be expressed in terms of the height-height correlation function,

$$C_c(k_\perp, \mathbf{r}) = e^{-\frac{1}{2}(k_\perp)^2 H_c(\mathbf{r})} . \qquad (2.19)$$

§II.2.2 *Height difference function of a rough crystalline surface*

In a crystalline surface, calculation of the height difference function requires the consideration of the discrete nature of the multilevel stepped structure. We have discussed in section §II.1.2 the contribution of the lateral surface feature to the diffraction structure factor from a discrete surface. The contribution of the vertical surface feature to the height difference function, (and then the structure factor) from a discrete surface, will be discussed next.

In contrast to the continuous surface model where a continuous Gaussian distribution is assumed, a discrete version of the Gaussian distribution function has to be used for a crystalline surface in which, $\Delta z = c\Delta h = 0, \pm c, \pm 2c, \pm 3c, \ldots$ The discrete version of Eq. (2.17) that gives a height difference function is expressed by

$$C_d(k_\perp, \mathbf{r}) = \langle e^{ik_\perp c\,[h(\mathbf{r}) - h(0)]} \rangle$$

$$= \sum_{m=-\infty}^{+\infty} g(\Delta z = mc, \mathbf{r}) e^{ik_\perp c\, m} . \qquad (2.17')$$

We can assume a discrete Gaussian distribution, $g(\Delta z = mc\,,\, \mathbf{r}) = C^{-1} e^{-\frac{m^2}{2\sigma^2}}$, where C is the normalization constant, $C = \sum_{m=-\infty}^{+\infty} e^{-\frac{m^2}{2\sigma^2}}$. For simplicity, we define the phase, $\phi = k_\perp c$. Equation (2.17') then becomes

$$C_d(k_\perp, \mathbf{r}) = \langle e^{\,i\phi\,[h(\mathbf{r}) - h(0)]} \rangle$$

$$= C^{-1} \sum_{m=-\infty}^{+\infty} e^{-\frac{m^2}{2\sigma^2}} e^{im\phi} \,. \qquad (2.20)$$

In order to analyze $C_d(k_\perp, \mathbf{r})$, one can employ an identity,[2.6]

$$f(z') = \sum_{m=-\infty}^{+\infty} \delta(z' - m) = \sum_{n=-\infty}^{+\infty} e^{-i2n\pi z'} \,,$$

where $f(z')$ is a one-dimensional lattice function. This function has a form similar to the two-dimensional lattice function shown in Eqs. (2.5) and (2.12). Equation (2.20) can then be written as

$$C(k_\perp, \mathbf{r}) = C^{-1} \int_{-\infty}^{+\infty} dz'\, e^{-\frac{z'^2}{2\sigma^2}} e^{iz'\phi} \sum_{m=-\infty}^{+\infty} \delta(z' - m)$$

$$= C^{-1} \sum_{n=-\infty}^{+\infty} \int_{-\infty}^{+\infty} dz'\, e^{-\frac{z'^2}{2\sigma^2}} e^{iz'(\phi - 2n\pi)}$$

$$= \sigma\sqrt{2\pi}\, C^{-1} \sum_{n=-\infty}^{+\infty} e^{-\frac{1}{2}(\phi - 2n\pi)^2 \sigma^2} \,,$$

with the normalization constant,

$$C = \sum_{m=-\infty}^{+\infty} e^{-\frac{m^2}{2\sigma^2}}$$

$$= \int_{-\infty}^{+\infty} dz' \, e^{-\frac{z'^2}{2\sigma^2}} \sum_{m=-\infty}^{+\infty} \delta(z' - m)$$

$$= \sum_{n=-\infty}^{+\infty} \int_{-\infty}^{+\infty} dz' \, e^{-\frac{z'^2}{2\sigma^2}} e^{i2n\pi z'}$$

$$= \sigma\sqrt{2\pi} \sum_{n=-\infty}^{+\infty} e^{-\frac{1}{2}(2n\pi)^2 \sigma^2} .$$

A numerical calculation indicates that for $\sigma > 0.8$, $\sigma^2 \approx C^{-1} \sum_{m=-\infty}^{+\infty} m^2 \, e^{-\frac{m^2}{2\sigma^2}}$. For example, if $\sigma = 0.8$, $C^{-1} \sum_{m=-\infty}^{+\infty} m^2 e^{-\frac{m^2}{2\sigma^2}} \approx 0.9998\sigma^2$ and if $\sigma = 1$, $C^{-1} \sum_{m=-\infty}^{+\infty} m^2 e^{-\frac{m^2}{2\sigma^2}} \approx 0.9999998\sigma^2$. Therefore, as long as $\sigma > 0.8$, we have the relation,

$$H_d(\mathbf{r}) = \langle [h(\mathbf{r}) - h(0)]^2 \rangle \approx \sigma^2 \approx C^{-1} \sum_{m=-\infty}^{+\infty} m^2 \, e^{-\frac{m^2}{2\sigma^2}}, \qquad (2.21)$$

which is similar to Eq. (2.18). Note that the height-height correlation function defined for a discrete surface, given by Eq. (2.21), has different units than that of a continuous surface, as given by Eq. (2.18). One has to multiply Eq. (2.21) by c^2 in order to compare the two equations.

Combining all of the above results, we finally obtain an explicit form for the height difference function in a crystalline surface,

$$C_d(k_\perp, \mathbf{r}) \approx \frac{\sum_{n=-\infty}^{+\infty} e^{-\frac{1}{2}H_d(\mathbf{r})(\phi - 2\pi n)^2}}{\sum_{n=-\infty}^{+\infty} e^{-\frac{1}{2}H_d(\mathbf{r})(2\pi n)^2}} . \tag{2.22}$$

Note that this result is quite different from the form generated from a continuous surface given by Eq. (2.19). The summation in Eq. (2.22) arises from the discrete nature of the lattice structure in a crystalline surface. We now discuss several very important aspects of the height difference function given by Eq. (2.22).

(i) Periodic oscillation

As an immediate result due to the periodic and discrete structure of a crystalline surface, Eq. (2.22) shows the reciprocal space ($k_\perp$-space) periodic relations, $C_d(k_\perp, \mathbf{r}) = C_d(k_\perp + \frac{2m\pi}{c}, \mathbf{r})$ and $C_d(\frac{2m\pi}{c}, \mathbf{r}) = 1$, $m = 0, 1, 2, ...$ Since the relative height, $[h(\mathbf{r}) - h(0)]$, can only have integer values, the phase term in Eq. (2.20) must have the relations, $e^{i(\phi+2n\pi)[h(\mathbf{r}) - h(0)]} = e^{i\phi[h(\mathbf{r}) - h(0)]}$ and $e^{i2n\pi[h(\mathbf{r}) - h(0)]} = 1$. Such relations do not exist in the case of the continuous surface model. To show the difference of the height difference functions between a crystalline surface and a continuous surface, we plot Eq. (2.22) and Eq. (2.19) as a function of $k_\perp c/\pi$ in Fig. 2.4(a) and Fig. 2.4(b) respectively, with a fixed parameter, $H_c(\mathbf{r})/c^2 = H_d(\mathbf{r}) = \sigma^2/c^2 = 0.5$. The continuous height difference function of Eq. (2.19) monotonically decays to zero as $k_\perp$ increases (Fig. 2.4(b)). In contrast, the discrete height difference function Eq. (2.22) oscillates periodically with a periodicity of $2\pi/c$ (Fig. 2.4(a)).

(ii) Lowest order approximation

In Eq. (2.22), if $H_d(\mathbf{r}) > 1$, the denominator, $\sum_{n=-\infty}^{+\infty} e^{-\frac{1}{2}H_d(\mathbf{r})(2\pi n)^2} \approx 1$, because $e^{-\frac{1}{2}H_d(\mathbf{r})(2\pi n)^2} \to 0$ for any $n \neq 0$. On the other hand, among the terms, $e^{-\frac{1}{2}H_d(\mathbf{r})(\phi - 2\pi n)^2}$, in the numerator of Eq. (2.22), the maximum one is such a term in which $|\phi - 2\pi n|$ has a minimum value, $|\phi - 2\pi n|_{min}$. In order to calculate $|\phi - 2\pi n|_{min}$, one defines a symbol: "[ ]", in which $[\phi]$ means ϕ modulo 2π such that $-\pi \leq [\phi] \leq \pi$.[2.7] One can easily show that $|\phi - 2\pi n|_{min} = |[\phi]|$. For example, if $\phi = 6\frac{1}{3}\pi$, we have

$$|\phi - 2\pi n|_{min} = \underset{-\infty < n < +\infty}{\mathrm{MIN}} \quad |6\frac{1}{3}\pi - 2\pi n| = \frac{1}{3}\pi,$$

which is consistent with the result, $[6\frac{1}{3}\pi] = \frac{1}{3}\pi$.

Besides the maximum term shown above, we can also find the next largest term, which is $e^{-\frac{1}{2}(2\pi - |[\phi]|)^2 H_d(\mathbf{r})}$. If the value of ϕ is not close to $(2n-1)\pi$, for example, $|[\phi]| \leq 0.5\pi$, the inequality, $e^{-\frac{1}{2}[\phi]^2 H_d(\mathbf{r})} >> e^{-\frac{1}{2}(2\pi - |[\phi]|)^2 H_d(\mathbf{r})}$, is valid. This inequality indicates that the summation in the numerator of Eq. (2.22), $\sum_{n=-\infty}^{+\infty} e^{-\frac{1}{2}H_d(\mathbf{r})(\phi - 2\pi n)^2}$, can always be represented only by its maximum term as long as the diffraction phase ϕ is not close to $(2n-1)\pi$. The contributions from the rest of the terms are negligibly small because for any $(\phi - 2\pi n) \neq [\phi]$, we have

$$e^{-\frac{1}{2}[\phi]^2 H_d(\mathbf{r})} >> e^{-\frac{1}{2}H_d(\mathbf{r})(\phi - 2\pi n)^2} \to 0.$$

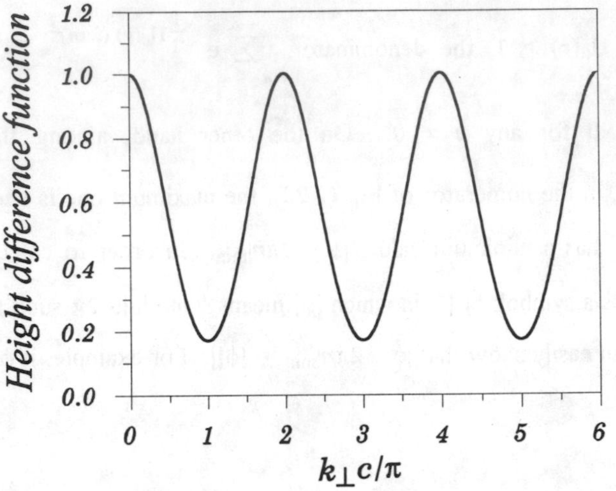

Fig. 2.4(a) Height difference function $C_d(k_\perp,\mathbf{r})$ vs. $k_\perp$ from a crystalline surface for $H_d(\mathbf{r}) = 0.5$.

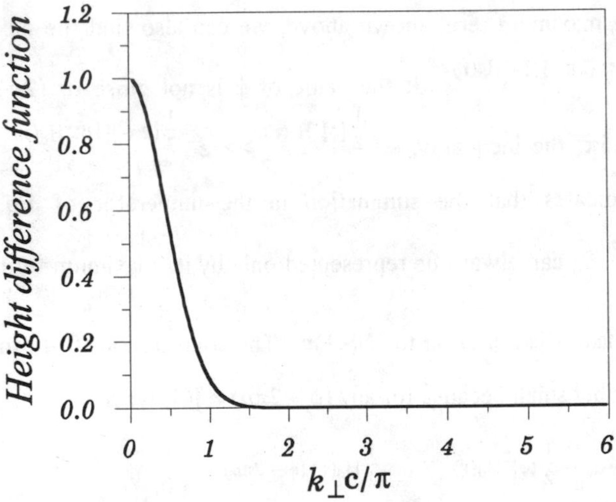

Fig. 2.4(b) Height difference function $C_c(k_\perp,\mathbf{r})$ vs. $k_\perp$ from a continuous surface for $H_c(\mathbf{r})/c^2 = 0.5$. In order to compare with Fig. 2.4(a), we use the same units, (c/π), for $k_\perp$, as used in Fig. 2.4(a).

In conclusion, the lowest order approximation for Eq. (2.22) can be represented as

$$C_d(\, k_\perp, \mathbf{r}\,) \approx e^{-\frac{1}{2}[\phi]^2 H_d(\mathbf{r})}, \qquad (\,|[\phi]| \neq \pi\,), \qquad (\,2.23\,)$$

which is valid under the condition that $H_d(\mathbf{r}) \geq 1$ and ϕ is not close to $(2n-1)\pi$. The diffraction condition, $\phi = (2n-1)\pi$, corresponds to the out-of-phase condition where a destructive interference occurs between terraces separated by an atomic step.

As a comparison, we plot respectively the rigorous height difference function, Eq. (2.22), and the lowest order approximation, Eq. (2.23), as a function of $\phi = k_\perp c$ in Fig. 2.5, where $H_d(\,\mathbf{r}\,) = 0.5$. It is shown that Eq. (2.23) agrees well with Eq. (2.22) at the vicinity of $\phi = 2n\pi$ but not at the vicinity of the out-of-phase condition, $\phi = (2n-1)\pi$. The diffraction condition, $\phi = 2n\pi$, is called the in-phase condition at which constructive interference occurs during the diffraction.

(iii) In-phase and out-of-phase diffraction conditions

In Figs. 2.4(a) and 2.4(b), the plots indicate that in the small ϕ regime, $0 \leq \phi = k_\perp c << \pi$, corresponding to the small $k_\perp$ diffraction condition, the height difference function of a crystalline surface, Eq. (2.22), (see Fig. 2.4(a)), is very close to that of a continuous surface, Eq. (2.19), (see Fig. 2.4(b)). This can also be seen analytically in Eq. (2.23) where if $0 \leq \phi << \pi$, we have $[\phi] = \phi$, so that

$$C_o(\, k_\perp, \mathbf{r}\,) \approx e^{-\frac{1}{2}\phi^2 H_d(\mathbf{r})}$$

$$= e^{-\frac{1}{2}(k_\perp)^2 H_d(\mathbf{r})\, c^2}, \qquad (\,0 \leq \phi << \pi\,),$$

which is identical to Eq. (2.19) for the continuous surface model if we assume an equal height-height correlation function, $H_c(\mathbf{r}) = H_d(\mathbf{r})\, c^2$.

The fact that Eq. (2.22) approximates Eq. (2.19) at $0 \le \phi << \pi$ has a simple physical meaning. Small $k_\perp$ corresponds to the long wave length diffraction condition in which the discrete atomic effect of a surface is negligibly small. The continuous surface model description is thus a good approximation at $0 \le \phi << \pi$.

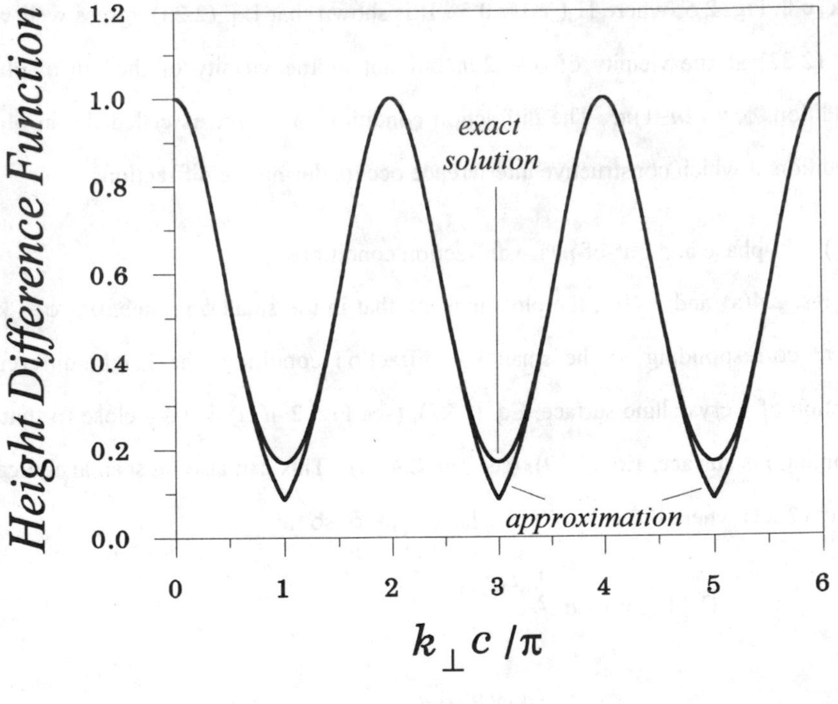

Fig. 2.5 Height difference functions from a crystalline surface: the exact solution (Eq. 2.22) and the lowest order approximation (Eq. 2.23).

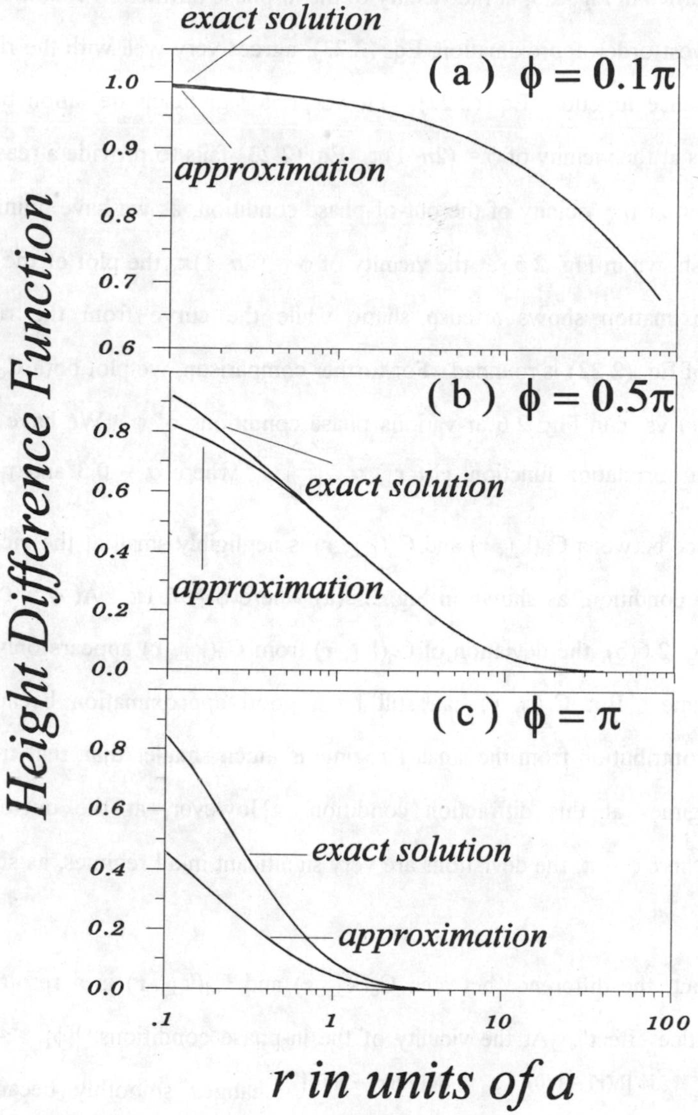

Fig. 2.6 Height difference function vs. r from a crystalline surface at: (a) the near in-phase condition $\phi = k_\perp c = 0.1\pi$; (b) $\phi = 0.5\pi$; (c) the out-of-phase condition, $\phi = \pi$. The calculations are based on the exact solution (Eq. 2.22) and the lowest order approximation (Eq. 2.23).

As shown in Fig. 2.5, at the vicinity of the in-phase diffraction conditions, $\phi = 2n\pi$, the lowest order approximation, Eq. (2.23), agrees very well with the rigorous height difference function, Eq. (2.22). However, a significant deviation between them appears at the vicinity of $\phi = (2n-1)\pi$. Eq. (2.23) fails to provide a reasonable approximation at the vicinity of the out-of-phase condition, as we have pointed out earlier. As shown in Fig. 2.5, at the vicinity of $\phi = (2n-1)\pi$, the plot of the lowest order approximation shows a cusp shape while the curve from the rigorous calculation of Eq. (2.22) is rounded. For further comparison, we plot both $C_o(k_\perp, \mathbf{r})$ and $C_d(k_\perp, \mathbf{r})$ vs. r in Fig. 2.6 at various phase conditions of ϕ. We have used a height-height correlation function, $H_d(\mathbf{r}) \propto 2\left(\frac{r}{\eta}\right)^{2\alpha}$, where $\alpha = 0.3$ and $\eta = 7.0$. The difference between $C_o(k_\perp, \mathbf{r})$ and $C_d(k_\perp, \mathbf{r})$ is negligibly small at the vicinity of the in-phase condition, as shown in Fig. 2.6(a) where $\phi = 0.1\pi$. At $\phi = 0.5\pi$, as shown in Fig. 2.6(b), the deviation of $C_o(k_\perp, \mathbf{r})$ from $C_d(k_\perp, \mathbf{r})$ appears only in the small r regime. But $C_o(\phi, \mathbf{r})$ can still be a good approximation because the diffraction contribution from the small r regime is much smaller than that from the larger r regimes at this diffraction condition. However, at the out-of-phase condition, where $\phi = \pi$, the deviations are very significant in all regimes, as shown in Fig. 2.6(c).

In fact, the difference between $C_o(k_\perp, \mathbf{r})$ and $C_d(k_\perp, \mathbf{r})$ is a result of the "discrete lattice effect". At the vicinity of the in-phase conditions, $|[\phi]| \ll \pi$, the phase term, $e^{i\phi\,[h(\mathbf{r}) - h(0)]} = e^{i[\phi]\,[h(\mathbf{r}) - h(0)]}$, changes smoothly because the variation of the phase factor $[\phi][h(\mathbf{r}) - h(0)]$ is very small even though the relative height, $[h(\mathbf{r}) - h(0)]$, can jump up or down by integral values. The situation where the phase factor $[\phi][h(\mathbf{r}) - h(0)]$ has small variations is quite similar to that in the "small $k_\perp$" condition for the continuous surface model. At the near in-phase condition, the phase term $e^{i[\phi]\,[h(\mathbf{r}) - h(0)]}$ is not sensitive to the discrete lattice effect which, therefore, can be ignored. The height difference function thus becomes the same as that in the continuous surface model. In contrast, at the vicinity of the out-of-phase conditions, $\|[\phi]\| \sim \pi$, the phase term, $e^{i\phi\,[h(\mathbf{r}) - h(0)]} \approx e^{i\pi\,[h(\mathbf{r}) - h(0)]}$, can change quite dramatically between −1 and +1 as the result of the discrete variation of the relative height $[h(\mathbf{r}) - h(0)]$. The origin of the dramatic fluctuation between −1 and +1 in the phase term comes from the destructive interference between terraces in the diffraction problem. Because of this destructive interference, the discrete lattice effect comes into play so that the lowest order term, $C_0(k_\perp, \mathbf{r})$, is not sufficient to represent $C_d(k_\perp, \mathbf{r})$. For $\|[\phi]\| \sim \pi$, both the maximum term $e^{-\frac{1}{2}[\phi]^2 H_d(\mathbf{r})}$ and the next largest term $e^{-\frac{1}{2}(2\pi - \|[\phi]\|)^2 H_d(\mathbf{r})}$ have about equal contribution to Eq. (2.22). The approximation, $C_0(k_\perp, \mathbf{r})$, which only contains the term $e^{-\frac{1}{2}[\phi]^2 H_d(\mathbf{r})}$, seemingly underestimates the value of $C_d(k_\perp, \mathbf{r})$. Therefore, at the vicinity of the out-of-phase condition, if $H_d(\mathbf{r}) \geq 1$, the lowest order of $C_d(k_\perp, \mathbf{r})$ should be expressed as

$$C_0(k_\perp, \mathbf{r}) \approx e^{-\frac{1}{2}[\phi]^2 H_d(\mathbf{r})} + e^{-\frac{1}{2}(2\pi - \|[\phi]\|)^2 H_d(\mathbf{r})}, \quad (\|[\phi]\| \sim \pi). \qquad (2.23')$$

§II.3 Height-Height Correlation Functions from Rough Surfaces

We have seen that the height difference function has an intrinsic connection to the statistical height-height correlation, as shown in both the continuous surface model, Eq. (2.19), and the crystalline surface model, Eq. (2.22). The height-height correlation function given by Eq. (2.18) is perhaps the most important and also the simplest correlation function defined to characterize the rough surface. It is much easier to understand the surface behavior from the height-height correlation function than that from the height difference function. In this section, we shall demonstrate the important aspects of the height-height correlation function in rough surfaces with non-divergent height fluctuations.

§II.3.1 *Correlation lengths and roughness parameter*

Normally, the description of a surface morphology requires a topography function, $z = z(x, y)$, which involves a huge number of parameters or degrees of freedom. However, because of the randomness, the parameters used for describing the surface morphology can be significantly reduced and the characterization of the complicated rough surface then becomes relatively simple. Instead of using the extremely complicated topography function, $z = z(x, y)$, the rough surface can be described statistically by a simple height-height correlation function. The height-height correlation function from a rough surface contains at least three important parameters. They are the vertical correlation length w, the lateral correlation length ξ and the roughness exponent α.

In order to characterize the surface roughness, w (the vertical correlation length) is a necessary parameter. w is defined as the root mean square surface height fluctuation as given by

$$w^2 = < [z(\mathbf{r}) - < z >]^2 >,$$

where $< z >$ is the average height of the surface. w is also called the interface width. This parameter describes the thickness of the solid-gas interface.

Many rough surfaces in nature exhibit non-diverging multilevel height fluctuations. The non-diverging height fluctuation is a common phenomenon during far-from-equilibrium film growth, where the surfaces are very rough but have a finite solid-gas interface width even when the surface area becomes very large ($\rightarrow \infty$). Such a morphology has been demonstrated in the surface topography of the Au:Pd thin film deposited on mica, as shown in Fig. 2.2. The Au:Pd film consists of many randomly distributed mountains and valleys. w thus provides an average measure of the mountain height or the valley depth.

The interface width w only characterizes the surface roughness along the vertical direction. However, the fluctuation of the surface level should also have intrinsic characteristics along the lateral directions. The lateral correlation length ξ is such a parameter that gives an average measure of the lateral characteristics. The height fluctuation could be considered as a transverse wave. w would be the amplitude of the fluctuation normal to the surface. ξ would be the wavelength of the fluctuations to characterize the spatial variation along the lateral surface direction. We also recall the real surface morphology shown in Fig. 2.2(b), where one not only needs a vertical parameter w to describe the average mountain height but also needs a parameter ξ to measure the average lateral size of the mountains or islands. The lateral correlation length ξ is thus the distance within which the surface variations are correlated but beyond which the surface fluctuations spread and are not correlated.

Based on the concept of the correlation lengths introduced above, we can derive a generic height-height correlation function on the long-range scale. According to the definition, we expand the height-height correlation function as

$$H(\mathbf{r}) = < [z(\mathbf{r}) - z(0)]^2 >$$

$$= < \{[z(\mathbf{r}) - <z>] - [z(0) - <z>]\}^2 >$$

$$= \langle [z(\mathbf{r}) - \langle z\rangle]^2 \rangle + \langle [z(0) - \langle z\rangle]^2 \rangle - 2\langle [z(\mathbf{r}) - \langle z\rangle][z(0) - \langle z\rangle] \rangle .$$

Over a distance much larger than the lateral correlation length, the surface height fluctuations should not be correlated and therefore,

$$\langle [z(\mathbf{r}) - \langle z\rangle][z(0) - \langle z\rangle] \rangle =$$

$$\approx \langle [z(\mathbf{r}) - \langle z\rangle] \rangle \langle [z(0) - \langle z\rangle] \rangle = 0.$$

The height-height correlation function thus has a long-range asymptotic form,

$$H(\mathbf{r}) = \langle [z(\mathbf{r}) - z(0)]^2 \rangle \sim 2\,w^2, \qquad (r \gg \xi), \qquad (2.24)$$

where we have use the identity,

$$\langle [z(\mathbf{r}) - \langle z\rangle]^2 \rangle = \langle [z(0) - \langle z\rangle]^2 \rangle = w^2.$$

The long-range height-height correlation reaches a constant value determined only by the "global" roughness w.

For the Au:Pd thin film shown in Fig. 2.2, the surface height fluctuation has a finite amplitude. The root mean square interface width of the topography has been measured to be w ~ 15.0 Å. The corresponding height-height correlation function $H_c(\mathbf{r})$ is plotted in Fig. 2.7. As shown in Fig. 2.7, $H_c(\mathbf{r})$ first increases with r in small r regime and then reaches a constant at the large r regime, r > 100 Å. This constant has been estimated to be ~ 450.0 Å, which is close to the value of $2w^2$. The lateral correlation length ξ is therefore ~ 100 Å.

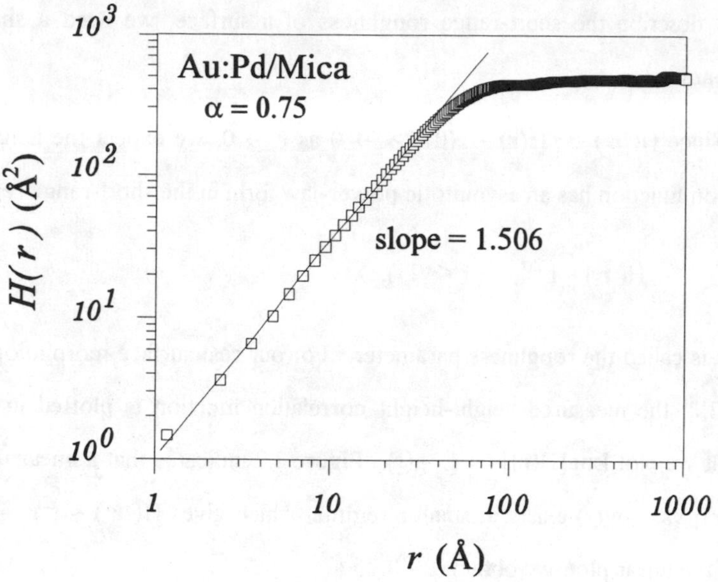

Fig. 2.7 The height-height correlation function measured from the Au:Pd thin film shown in Fig. 2.2.

So far, we have introduced two parameters to describe the global behavior in a rough surface. However, we have not described the detailed surface structure on a relatively short range scale. We must emphasize that even with the same w and ξ, the detailed local surface morphology for different systems can be completely different. One example is shown in Fig. 2.8 [2.8] in which we plot the surface profiles of different characteristics with the same interface width $w = 1.1 \pm 0.1$. Figure 2.8(c) exhibits a surface that has a jagged local structure, while the surface in Fig. 2.8(a) has a well-correlated and smooth-textured local profile. It seems clearly that the long-range parameters, w and ξ, are not sufficient to characterize a rough surface. In order to describe the short-range roughness of a surface, we need a short-range related parameter.

Since H(**r**) = <[z(**r**) − z(0)]^2> → 0 as **r** → 0, we expect the height-height correlation function has an asymptotic power-law form in the short-range regime,

$$H(\mathbf{r}) \sim r^{2\alpha}, \quad (r << \xi), \tag{2.25}$$

where α is called the roughness parameter. For our real surface morphology shown in Fig. 2.2, the measured height-height correlation function is plotted in Fig. 2.7. Note that we plot Log[H(**r**)] vs. Log(r). Figure 2.7 indicates that a linear relation of Log[H(**r**)] vs. Log(r) exists in small r regime, which gives H(**r**) ~ $r^{2\alpha}$. From the slope of the linear plot, we obtain α ~ 0.75.

The power-law form of the height-height correlation has been successfully used to describe the self-affine fractal surface morphology. Fractal surface is a common rough surface which can be found in many places in nature.[2.9] We will give a detailed discussion on the self-affine surface structure in Chapter IV.

The roughness exponent α can have a range, 0 ≤ α ≤ 1, for a self-affine fractal surface. α provides a measure of the surface roughness. Shown in Fig. 2.8 as an example,[2.8] a surface profile exhibited in Fig. 2.8(a) has a value of α = 0.7 while the morphology in Fig. 2.8(c) gives α = 0.3. Usually, a larger value of α (> 0.5) corresponds to a smooth-textured surface structure in the short-range while the smaller value of α (< 0.5) corresponds to a more jagged local surface morphology.[2.10]

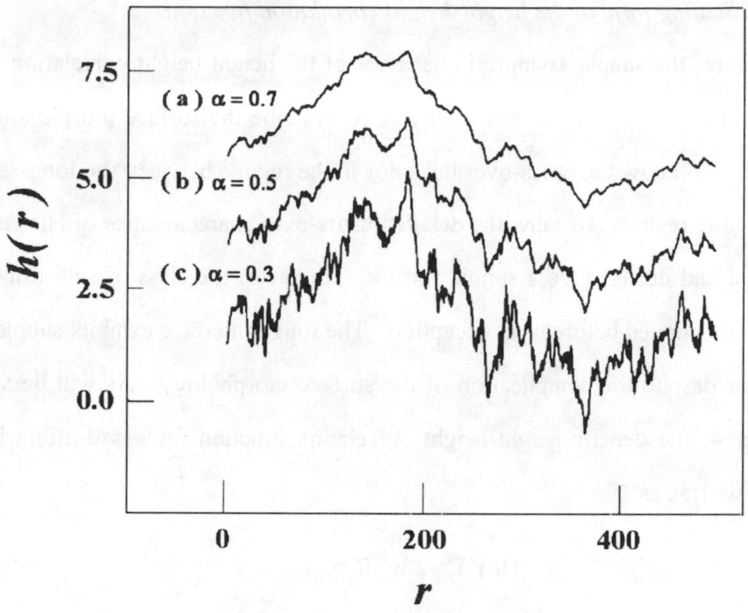

Fig. 2.8 Short range surface profiles at different values of α. These surfaces possess the same value of w and ξ, but different values of the roughness parameter α. The scales are in arbitrary units (from Ref. 2.8, courtesy of J. Krim).

In conclusion, a rough surface with a finite interface width can be described by the non-diverging surface height-height correlation function. The asymptotic forms of the height-height correlation on both the short-range and the long-range scales are given by

$$H(\mathbf{r}) \sim \begin{cases} r^{2\alpha} & \text{for } r \ll \xi, \\ 2w^2 & \text{for } r \gg \xi. \end{cases} \qquad (2.26)$$

H(**r**) is thus determined by the roughness exponent α, the vertical correlation length w and the lateral correlation length ξ.

§II.3.2 *Scaling form of the height-height correlation function*

Sometimes, the simple asymptotic behavior of the height-height correlation function shown in Eq. (2.26) is still not sufficient to describe the surface morphology. One may need to know the cross-over behavior in the regime between the long-range and short-range regime. Usually, the detailed cross-over characteristics of H(**r**) are more involved and do not give a simple form. However, the class of self-affine fractal surface mentioned before is an exception. The rough interface exhibits simple scaling behavior despite the complication of the surface morphology. As will be shown in Chapter 4, the generic height-height correlation function for a self-affine interface can be written as

$$H(\mathbf{r}) = 2\,w^2\, f(\frac{r}{\xi})\,, \tag{2.27}$$

where f(X) is called the scaling function. f(X) has the properties of f(X) = 1, for X >>1 and $f(X) = X^{2\alpha}$ for X << 1. These properties are consistent with both the short-range characteristic shown in Eq. (2.25) and the long-range asymptotic behavior given by Eq. (2.24). Between the asymptotic regimes, Eq. (2.27) also includes the description of the cross-over regime ($r \sim \xi$). One of the simplest phenomenological scaling function f(X) for a surface has been proposed by Sinha *et al.* to be [2.11]

$$f(X) = 1 - e^{-X^{2\alpha}}. \tag{2.28}$$

In Appendix (IIB), we show a one-dimensional Gaussian model, where a rigorous form solution for the height-height correlation function has been obtained as,

$$H(x) = 2\,w^2\,(1 - e^{-|x|/\xi}),$$

which gives a scaling function, $f(X) = 1 - e^{-X}$. In this model, $\alpha = 0.5$.

The scaling description in Eq. (2.27) leads to an explicit expression in the short-range regime,

$$H(\mathbf{r}) = 2\left(\frac{r}{\eta}\right)^{2\alpha}, \quad (r << \xi), \qquad (2.25')$$

where $\eta = \xi w^{-\frac{1}{\alpha}}$ is another very important parameter characterizing the short-range behavior in the rough surface. As we shall see in Chapter 6, the short-range parameter η in crystalline surfaces has a specific physical meaning: it is proportional to the average terrace size in the surface.[2.12]

From the meaning of η, the behavior of the height difference function in a crystalline rough surface can be understood more precisely. In Fig. 2.9, we plot a typical height difference function $C_d(k_\perp, \mathbf{r})$ in a crystalline surface, using Eq. (2.23) with $\phi = \pi$ and the height-height correlation function, $H_d(\mathbf{r}) = 2\,w^2\,(1 - e^{-|\mathbf{r}|/\xi})$, with $w = 1.5$, $\eta = 10$ and $\xi = \eta w^2 = 22.5$. As shown in Fig. 2.9, in the small r regime, $r << \eta$, $C_d(k_\perp, \mathbf{r}) \sim 1$, while as $r > \eta$, the function quickly decays to zero. The physical interpretation of this behavior is the following: the surface atoms at a distance, $r << \eta$, are located almost at same terraces so that the height difference, $\Delta h = h(\mathbf{r}) - h(0) \sim 0$, i.e., $e^{ik_\perp c\,\Delta h} = e^{i\pi\Delta h} \sim 1$. In contrast, for $r > \eta$, the atoms can locate at different terraces so that the surface height levels vary randomly among different levels. The uncertainty of the surface height among atoms at $r > \eta$ leads to a random fluctuation of the phase factor, $e^{ik_\perp c\,\Delta h} = e^{i\pi\Delta h}$, between the values, -1 and $+1$. Quantitatively, at large r, the fluctuation eventually leads to the vanishing of the average phase factor, $C_d(k_\perp, \mathbf{r}) = \langle e^{ik_\perp c\,\Delta h} \rangle \sim 0$.

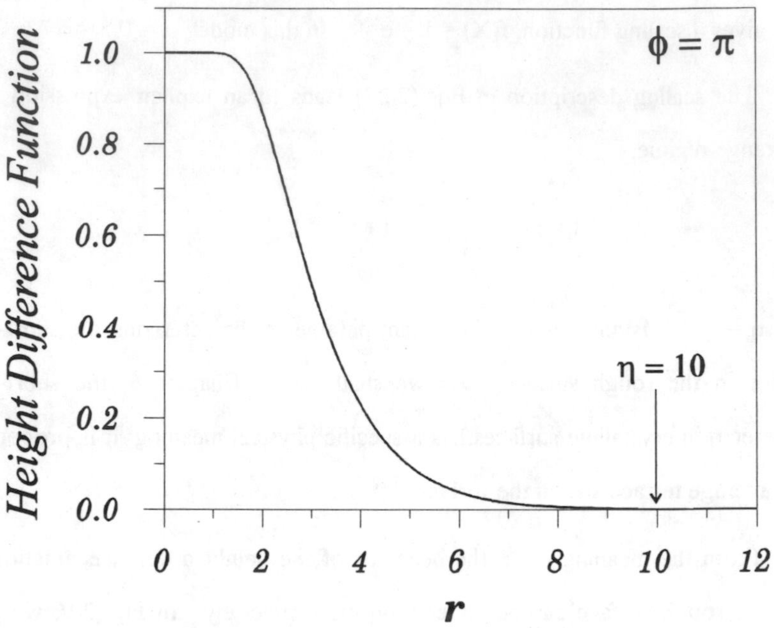

Fig. 2.9 Height difference function at the out-of-phase condition $\phi = \pi$ for a given average terrace size, $\eta = 10$. The lateral scales (for r and η) are in units of *a*.

In conclusion, based on the generic expression of the height-height correlation function given by Eq. (2.27), a rigorous and complete diffraction theory for characterizing rough surface morphologies can be formulated. The characterization of a rough surface in a diffraction experiment will concentrate on the measurement of w, ξ and α. The detailed diffraction theory will be discussed in Chapter 3.

REVIEW AND SUMMARY

Diffraction structure factor from statistically rough surfaces

Diffraction structure factor is a Fourier transform of the height difference function $C(k_\perp, \mathbf{r}) = \langle e^{ik_\perp[z(\mathbf{r}) - z(0)]} \rangle$,

$$S(\mathbf{k}) = \int d^2r\, C(k_\perp, \mathbf{r})\, e^{ik_\parallel \mathbf{r}} . \tag{2.16}$$

$k_\perp$ is the momentum transfer perpendicular to the surface.

The height difference function with a Gaussian Distribution: continuous surfaces

For statistically rough surfaces, the relative surface heights, $z(\mathbf{r}) - z(0)$, usually obey a statistical Gaussian distribution. The height difference function can therefore be derived as a Gaussian function,

$$C_c(k_\perp, \mathbf{r}) = \langle e^{ik_\perp[z(\mathbf{r}) - z(0)]} \rangle = e^{-\frac{1}{2}(k_\perp)^2 H_c(\mathbf{r})} . \tag{2.19}$$

$H_c(\mathbf{r}) = \langle [z(\mathbf{r}) - z(0)]^2 \rangle$ is the height-height correlation function for continuous surfaces.

The height difference function with a Gaussian Distribution: crystalline surfaces

The discrete Gaussian Distribution for a crystalline surface leads to a different form of the height difference function,

$$C_d(k_\perp, \mathbf{r}) = \frac{\sum_{n=-\infty}^{+\infty} e^{-\frac{1}{2} H_d(\mathbf{r}) (\phi - 2\pi n)^2}}{\sum_{n=-\infty}^{+\infty} e^{-\frac{1}{2} H_d(\mathbf{r}) (2\pi n)^2}} . \tag{2.22}$$

$H_d(\mathbf{r}) = \frac{1}{c^2} \langle [z(\mathbf{r}) - z(0)]^2 \rangle$ is the height-height correlation function for crystalline surfaces and c denotes the surface layer spacing. $\phi = k_\perp c$ determines the diffraction phase conditions. $\phi = k_\perp c = 2m\pi$ are called in-phase conditions and $\phi = k_\perp c = (2m-1)\pi$ are out-of-phase conditions, where $m = 0, 1, 2, \ldots$

Lowest order approximation of Eq. (2.22):

(1) For away from the out-of-phase condition, $|[\phi]| \neq \pi$,

$$C_d(k_\perp, \mathbf{r}) \approx e^{-\frac{1}{2}[\phi]^2 H_d(\mathbf{r})}. \qquad (2.23)$$

(2) For near the out-of-phase condition, $|[\phi]| \sim \pi$,

$$C_d(k_\perp, \mathbf{r}) \approx e^{-\frac{1}{2}[\phi]^2 H_d(\mathbf{r})} + e^{-\frac{1}{2}(2\pi - |[\phi]|)^2 H_d(\mathbf{r})}, \qquad (2.23')$$

where $[\phi]$ means ϕ modulo 2π such that $-\pi \leq [\phi] \leq \pi$.

Characterization of surface roughness

In order to characterize statistically rough surfaces, one needs at least three parameters:

1. *Lateral correlation length* ξ

ξ is a distance within which the surface variations are correlated but beyond which the surface fluctuations spread and are not correlated. ξ characterizes the spatial variation of a rough surface along the lateral direction.

2. *Interface width* w

$w = \sqrt{<[z(\mathbf{r}) - <z>]^2>}$ is a measure of roughness, or thickness of the solid-gas interface and characterizes the spatial variation of a rough surface along the vertical direction. Note that the w only describes the surface roughness in the long range ($> \xi$).

3. *Roughness exponent* α

α is an important parameter to describe the self-affine fractal surfaces. For surfaces having non-divergent interface width, α provides a measure of the surface roughness in the short range ($< \xi$). A larger value of α (> 0.5) corresponds to a smooth-textured, short-range surface structure while the smaller value of α (< 0.5) corresponds to a more jagged, local surface morphology.

Height-height correlation function

For a self-affine rough surface, the height-height correlation function can be expressed as

$$H(\mathbf{r}) = 2w^2 f(\frac{r}{\xi}) \sim \begin{cases} r^{2\alpha} & \text{for short range } (r < \xi)\,, \\ 2w^2 & \text{for long range } (r > \xi)\,, \end{cases} \tag{2.27}$$

where f(X) is called the scaling function.

Appendix IIA Proof of Eq. (2.12)

Consider a finite two-dimensional reciprocal lattice with reciprocal unit vectors, $\boldsymbol{a}^*$ and $\boldsymbol{b}^*$, which have been defined in §I.2.3. For simplicity, we assume $\boldsymbol{a}^* \perp \boldsymbol{b}^*$ with $\boldsymbol{a}^*$ and $\boldsymbol{b}^*$ along x and y directions respectively. The reciprocal lattice vector is then given by $G_{kl} = k\,\boldsymbol{a}^* + l\,\boldsymbol{b}^* = k\,a^*\,\mathbf{e}_x + l\,b^*\,\mathbf{e}_y$, where $a^*=2\pi/a$ and $b^*=2\pi/b$. The finite lattice size is assumed to be

$$|(2N+1)\boldsymbol{a}^* \times (2N+1)\boldsymbol{b}^*| = (2N+1)^2|\boldsymbol{a}^* \times \boldsymbol{b}^*| = (2N+1)^2\, a^*\, b^* = (2\pi)^2\,(2N+1)^2\,v^{-1},$$

where $v = |\boldsymbol{a} \times \boldsymbol{b}| = ab$ denotes the area of the real-space lattice unit mesh.

Letting $\boldsymbol{\rho} = x\mathbf{e}_x + y\mathbf{e}_y$, we calculate the following summation,

$$\frac{1}{v}\sum_{k,\,l=-N}^{N} e^{-iG_{kl}\cdot\rho} = \sum_{k=-N}^{N} e^{-ik\boldsymbol{a}^*\cdot\rho} \times \sum_{l=-N}^{N} e^{-il\boldsymbol{b}^*\cdot\rho}$$

$$= \frac{1}{v}\,\frac{\sin[(N+1/2)\rho.\boldsymbol{a}^*]}{\sin(\rho.\boldsymbol{a}^*/2)} \times \frac{\sin[(N+1/2)\rho.\boldsymbol{b}^*]}{\sin(\rho.\boldsymbol{b}^*/2)}$$

$$= \frac{1}{v}\,\frac{\sin[(2N+1)\pi x/a]}{\sin(\pi x/a)} \times \frac{\sin[(2N+1)\pi y/b]}{\sin(\pi y/b)}.$$

This summation can reduce to the right hand side of Eq. (2.12), $\frac{1}{v}\sum_{k,\,l} e^{-iG_{kl}\cdot\rho}$, when $N\rightarrow\infty$.

Using the following identity,

$$\frac{\sin(Mz)}{\sin(z)} \sim \sum_{m} \pi\,\delta(\,z - m\pi\,), \quad (\text{as } M \rightarrow \infty),$$

where $m = 0, \pm 1, \pm 2, \ldots$, we can show that

$$\frac{1}{V}\sum_{k,\,l} e^{-i\mathbf{G}_{kl}\cdot\boldsymbol{\rho}} =$$

$$= \lim_{N \to +\infty} \frac{1}{V}\sum_{k,\,l=-N}^{N} e^{-i\mathbf{G}_{kl}\cdot\boldsymbol{\rho}} = \frac{1}{V}\sum_{m} \pi\,\delta(\,\pi x/a - m\pi\,) \sum_{n} \pi\delta(\,\pi y/b - n\pi).$$

Using the identity, $\delta(\,C\,z\,) = C^{-1}\,\delta(\,z\,)$, we obtain $\delta(\pi x/a - m\pi) = (a/\pi)\,\delta(x - ma)$ and $\delta(\pi y/b - n\pi\,) = (b/\pi)\,\delta(y - nb)$. The above equation can thus be rewritten as

$$\frac{1}{V}\sum_{k,\,l} e^{-i\mathbf{G}_{kl}\cdot\boldsymbol{\rho}} = \sum_{m} \delta(x - ma) \sum_{n} \delta(y - nb) = \sum_{m,\,n} \delta(\,\boldsymbol{\rho} - \boldsymbol{\rho}_{mn}\,),$$

which is just Eq. (2.12).

Appendix IIB One-dimensional Gaussian Model Surface

In order to describe statistically a non-divergent rough surface morphology, we propose a solvable one-dimensional surface model which captures the essential physics in this kind of surfaces.

Consider a one-dimensional surface which has a discrete lattice structure in the lateral direction (x-direction), but vertically (z-direction), the surface height can change continuously.

Define $f(x, z_0, z)$ as the probability density that a surface atom is found at the site (x, z) when a given surface atom exists at the position $(0, z_0)$. $f(x, z_0, z)$ is also defined as the pair correlation function between the two surface positions, (x, z) and $(0, z_0)$. Both z and z_0 take continuous values but x can only take discrete value, $x = 0, \pm1, \pm2, \ldots$ Based on the definition, we can have the following relation,

$$f(x+1, z_0, z) = \int_{-\infty}^{+\infty} dz' \, f(x, z_0, z') f(1, z', z) \, . \qquad (\text{ApIIB.1})$$

In order to obtain $f(x, z_0, z)$ from $f(1, z', z)$, we consider the Fourier transform of the pair correlation function,

$$F(x, z_0, q) = \frac{1}{\sqrt{2\pi}} \int_{-\infty}^{+\infty} dz \, f(x, z_0, z) \, e^{iqz} \, . \qquad (\text{ApIIB.2})$$

By Fourier transforming Eq. (ApIIB.1), we obtain

$$F(x+1, z_0, q) = \int_{-\infty}^{+\infty} dz' \, f(x, z_0, z') \frac{1}{\sqrt{2\pi}} \int_{-\infty}^{+\infty} dz \, f(1, z', z) \, e^{iqz}$$

$$= \int_{-\infty}^{+\infty} dz' \, f(x, z_0, z') \, F(1, z', q) \, . \qquad (\text{ApIIB.3})$$

Shown in Eqs. (ApIIB.1) and (ApIIB.3), the pair correlation $f(x, z_0, z)$ solely depends on the nearest neighbor probability function, $f(1, z', z)$, which is the probability density that a surface height is found to be z at the nearest neighbor site $x = 1$ when a given surface height at the site $x = 0$ is given to be z'. Considering two key ingredients in most rough surfaces, i.e., the Gaussian height distribution and the non-divergent height fluctuation, we assume that the probability $f(1, z', z)$ obeys a modified Gaussian distribution function,

$$f(1, z', z) \propto e^{-\kappa(z' - z)^2} e^{-\nu z^2} ,$$

where the term, $e^{-\kappa(z' - z)^2}$, represents the nearest-neighbor interaction and κ is a quantity related to the surface tension. This term shows that at the "local" regimes, the surface prefers at the same level and the relative height, $z - z'$, is distributed according to a Gaussian function. On the other hand, the term, $e^{-\nu z^2}$, represents a "global" interaction with a dimensionless "stabilizing field" ν. Such a "global" interaction tends to localize the surface height near $< z > = 0$ and therefore, can insure a finite height fluctuation in the surface.

Using the normalization condition, $\int_{-\infty}^{+\infty} dz \, f(1, z', z) = 1$, one can express the modified Gaussian distribution function $f(1, z', z)$ as

$$f(1, z', z) = \sqrt{\frac{\kappa + \nu}{\pi}} \exp[-(\kappa + \nu)(z - \frac{\kappa}{\kappa + \nu} z')^2] \, . \qquad (\text{ApIIB.4})$$

The Fourier transform of $f(1, z', z)$ is then given by

$$F(1, z', q) = \frac{1}{\sqrt{2\pi}} \int_{-\infty}^{+\infty} dz\, f(1, z', z)\, e^{iq\,z}$$

$$= \frac{1}{\sqrt{2\pi}} \exp[-\frac{q^2}{4(\kappa+\nu)}] \exp[iq\,z'\frac{\kappa}{\kappa+\nu}]. \qquad (\text{ApIIB.5})$$

Inserting Eq. (ApIIB.5) into Eq. (ApIIB.3), we obtain

$$F(x+1, z_o, q) = \frac{1}{\sqrt{2\pi}} \exp[-\frac{q^2}{4(\kappa+\nu)}] \int_{-\infty}^{+\infty} dz'\, f(x, z_o, z') \exp[iq\,z'\frac{\kappa}{\kappa+\nu}]$$

$$= \exp[-\frac{q^2}{4(\kappa+\nu)}]\, F(x, z_o, q\frac{\kappa}{\kappa+\nu}).$$

We can further rewrite the above equation as

$$F(x+1, z_o, q) = \exp[-\frac{q^2}{4(\kappa+\nu)}\gamma(x)]\, F(1, z_o, q\left(\frac{\kappa}{\kappa+\nu}\right)^x)$$

$$= \frac{1}{\sqrt{2\pi}} \exp[iq\,z_o\left(\frac{\kappa}{\kappa+\nu}\right)^{x+1}] \exp[-\frac{q^2}{4(\kappa+\nu)}\gamma(x+1)], \qquad (\text{ApIIB.6})$$

where

$$\gamma(x) = \sum_{j=0}^{x-1} \left(\frac{\kappa}{\kappa+\nu}\right)^{2j}$$

$$= [1-\left(\frac{\kappa}{\kappa+\nu}\right)^{2x}]\,[1-\left(\frac{\kappa}{\kappa+\nu}\right)^{2}]^{-1}$$

$$= \frac{(\kappa+\nu)^2}{\nu(2\kappa+\nu)}[1-\left(\frac{\kappa}{\kappa+\nu}\right)^{2x}]. \qquad (\text{ApIIB.7})$$

By Fourier transforming Eq. (ApIIB.6) back to real space, we obtain a closed form function for the pair correlation function,

$$f(x, z_o, z) = \frac{1}{\sqrt{2\pi}} \int_{-\infty}^{+\infty} dz\, F(x, z_o, q)\, e^{-iqz}$$

$$= \sqrt{\frac{\kappa+\nu}{\pi\gamma(x)}} \exp\left(-\frac{\kappa+\nu}{\gamma(x)}[z - z_o(\frac{\kappa}{\kappa+\nu})^x]^2\right). \qquad \text{(ApIIB.8)}$$

If $x = 0$, $\gamma(x) = 0$, then Eq. (ApIIB.8) becomes $f(0, z_o, z) = \delta(z - z_o)$, which is consistent with the definition of the pair correlation function.

On the other hand, if $x \to \infty$, $\gamma(\infty) = \frac{(\kappa+\nu)^2}{\nu(2\kappa+\nu)}$ and the pair correlation function turns out to be the absolute surface height distribution function, $\rho_o(z)$, i.e.,

$$\rho_o(z) = f(\infty, z_o, z) = \frac{1}{w\sqrt{2\pi}} \exp[-\frac{z^2}{2w^2}], \qquad \text{(ApIIB.9)}$$

where $w^2 = \frac{\kappa+\nu}{2\nu(2\kappa+\nu)}$. The average surface level can then be calculated as

$$< z > = \int_{-\infty}^{+\infty} z\, \rho_o(z)dz = 0,$$

with a standard deviation, $< [z - < z >]^2 > = \int_{-\infty}^{+\infty} z^2\, \rho_o(z)dz = w^2$. As we have expected, the surface height distribution is localized at the average surface level with a fluctuation amplitude ~ w. The root mean square height fluctuation w is also called as the interface width.

Combining Eqs. (ApIIB.8) with (ApIIB.9) and letting $\Delta z = z(x) - z(0) = z - z_o$, we can obtain the relative height distribution which has a rigorous Gaussian form function,

$$g(\Delta z, x) = \int_{-\infty}^{+\infty} dz_o\, \rho_o(z_o)\, f(x, z_o, z_o + \Delta z) = \frac{1}{\omega\sqrt{2\pi}} \exp[-\frac{(\Delta z)^2}{2\omega^2}], \quad \text{(ApIIB.10)}$$

where $\omega^2 = \frac{\kappa+\nu}{\nu(2\kappa+\nu)}\left(1 - (\frac{\kappa}{\kappa+\nu})^x\right)$. The height-height correlation function can then be given by $< [z(x) - z(0)]^2 > = \int_{-\infty}^{+\infty} d(\Delta z)\, \rho(\Delta z, x)\, (\Delta z)^2 = \omega^2$, i.e.,

$$H(x) = < [z(x) - z(0)]^2 > = 2\, w^2 (1 - e^{-x/\xi}), \quad \text{(ApIIB.11)}$$

where $\xi^{-1} = \mathrm{Ln}\left(\frac{\kappa+\nu}{\kappa}\right)$. ξ is called the correlation length.

REFERENCES

2.1 T.-M. Lu and M. G. Lagally, *Surf. Sci.* **120**, 47 (1982).

2.2 H.-N. Yang, A. Chan, and G.-C. Wang, *J. Appl. Phys.*, to be published.

2.3 M. W. Mitchell and D. A. Bonnell, *J. Mater. Res.* **5**, 2244 (1990).

2.4 G. Reiss, F. Schneider, J. Vancea, and H. Hoffman, *J. Appl. Phys.* **67**, 867 (1990).

2.5 G. Reiss, J. Vancea, J. Wittman, J. Zweck, and H. Hoffman, *J. Appl. Phys.* **67**, 1156 (1990).

2.6 J. Villain, D. R. Grempel, and J. Lapujoulade, *J. Phys.* **F15**, 809 (1985).

2.7 I. K. Robinson, E. H. Conrad, and D. S. Reed, *J. Phys.* (*Paris*) **51**, 103 (1990).

2.8 R. Chiarello, V. Panella, J. Krim, and C. Thompson, *Phys. Rev. Lett.* **67**, 3408 (1991).

2.9 B. B. Mandelbrot, *The Fractal Geometry of Nature* (Freeman, New York, 1982).

2.10 This statement is correct only from a short-range scale point of view, $L < \xi$. In contrast, from a long-range scale point of view, $L > \xi$, an increase in the roughness exponent α implies an increase in the perceived surface roughness. This scale-dependent problem has been pointed out recently by Krim and Indekeu (to be published in *Phys. Rev.* **E**, *Brief Reports*).

2.11 S. K. Sinha, E. B. Sirota, S. Garoff, and H. B. Stanley, *Phys. Rev.* **B38**, 2297 (1987).

2.12 H.-N. Yang, T.-M. Lu, and G.-C. Wang, *Phys. Rev. Lett.* **68**, 2612 (1992); Phys. Rev. **B47**, 3911 (1993).

Chapter III DIFFRACTION FROM ROUGH CRYSTALLINE AND NON-CRYSTALLINE SURFACES

With the specific height-height correlation function defined in the last Chapter, we now proceed to calculate the diffraction structure factor from rough crystalline and non-crystalline surfaces. It is shown that the diffraction structure factor in general contains a δ-function at the central peak position and a diffuse intensity which is a sum of an infinite number of a special function with different widths. [3.1] The differences between a crystalline and a non-crystalline structures are detailed. We also outline the methods of extracting the interface width w, the lateral correlation length ξ and the roughness parameter α from the diffraction structure factor.

§III.1 Diffraction Structure Factor from Rough Surfaces: General Approach

As we have shown in last chapter, an interesting and general class of height-height correlation function for rough surfaces with a non-divergent interface width has a scaling form described by Eq. (2.27). Combining this generic form of the height-height correlation function with the height difference function, Eq. (2.19), for a continuous surface or Eq. (2.22) for a crystalline surface, we are able to predict important aspects of the corresponding diffraction structure factor defined by Eq. (2.16).

§III.1.1 *General form of the diffraction structure factor*

We rewrite the height difference function, Eq. (2.19) or Eq. (2.22), as

$$C(k_\perp, \mathbf{r}) = C_\infty(k_\perp) + \Delta C(k_\perp, \mathbf{r}), \quad (3.1)$$

with

$$C_\infty(k_\perp) \equiv C(k_\perp, \mathbf{r} \to \infty),$$
$$\Delta C(k_\perp, \mathbf{r}) \equiv C(k_\perp, \mathbf{r}) - C(k_\perp, \mathbf{r} \to \infty)\,.$$

The height difference function is divided into the long-range part, $C_\infty(k_\perp)$, and the short-range part, $\Delta C(k_\perp, \mathbf{r})$. $\Delta C(k_\perp, \mathbf{r})$ approaches zero as $\mathbf{r} \to \infty$ while $C_\infty(k_\perp)$ is an **r**-independent constant. We can illustrate the behavior of Eq. (3.1) in Fig. 3.1, where we use the parameters, $w = 1.5$, $\alpha = 0.5$ and $\eta = 10.0$. Shown in Fig. 3.1(a) as the dashed straight line parallel to the r-axis, $C_\infty(k_\perp) = C(k_\perp, \mathbf{r} \to \infty)$ is a constant "background" throughout the entire real-space region between $r = 0$ and $r \to \infty$. $C_\infty(k_\perp)$ is thus related to the global (long-range) characteristics of the rough surface. In contrast, $\Delta C(k_\perp, \mathbf{r})$ is a function equal to $C(k_\perp, \mathbf{r})$ subtracted by the "background" $C_\infty(k_\perp)$, $\Delta C(k_\perp, \mathbf{r}) = C(k_\perp, \mathbf{r}) - C(k_\perp, \mathbf{r} \to \infty)$. As shown in Fig. 3.1(b), $\Delta C(k_\perp, \mathbf{r})$ quickly decays to zero at large r and therefore, is confined in the short-range regime. Figure 3.1(c) is the plot of $C(k_\perp, \mathbf{r})$ with a perpendicular wavevector, $k_\perp = \pi/c$, larger than that of Fig. 3.1(a) where $k_\perp = 0.3\pi/c$. For a large value of $k_\perp$, since the long-range term $C_\infty(k_\perp)$ could become negligibly small, as shown in Fig. 3.1(c), we have $\Delta C(k_\perp, \mathbf{r}) \approx C(k_\perp, \mathbf{r})$. Comparing Fig. 3.1(b) with Fig. 3.1(c), we notice that the extent of the short-range regime depends quite sensitively on $k_\perp$. For smaller $k_\perp$, the short-range regime is larger (Fig. 3.1(b)) while for larger $k_\perp$, as shown in Fig. 3.1(c), it becomes smaller. In the case of a crystalline surface, the range of the confined region for $\Delta C(k_\perp, \mathbf{r})$ can vary from a maximum ($\to \infty$) at near the in-phase diffraction condition to a minimum at the out-of-phase condition.

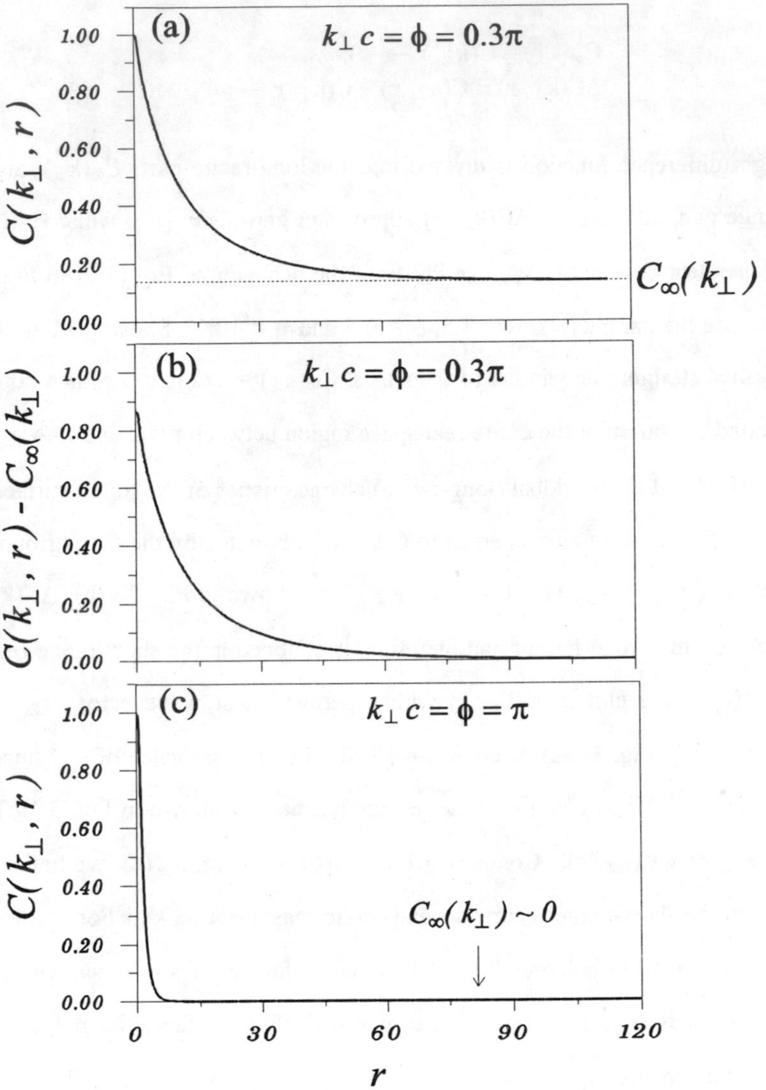

Fig. 3.1 Behaviors of the height difference functions of $C(k_\perp, \mathbf{r})$, $C_\infty(k_\perp)$ and $\Delta C(k_\perp, \mathbf{r})$. (a) $C(k_\perp, \mathbf{r})$ and $C_\infty(k_\perp)$ at $\phi = 0.3\pi$; (b) $\Delta C(k_\perp, \mathbf{r})$ at $\phi = 0.3\pi$; (c) $C(k_\perp, \mathbf{r})$ and $C_\infty(k_\perp)$ at $\phi = \pi$. The lateral scales are in units of a.

Accordingly, if we insert Eq. (3.1) into Eq. (2.16), the diffraction structure factor can also be divided into two parts: one is a sharp central δ-function associated with the long-range behavior; another one containing a broad diffuse component is connected to the short-range properties,

$$S(\mathbf{k}) = \int d^2r\, C_\infty(k_\perp)\, e^{i\mathbf{k}_\parallel\cdot\mathbf{r}} + \int d^2r\, \Delta C(k_\perp, \mathbf{r})\, e^{i\mathbf{k}_\parallel\cdot\mathbf{r}}$$

$$= (2\pi)^2 C_\infty(k_\perp)\, \delta(\mathbf{k}_\parallel) + S_{diff}(\mathbf{k}_\parallel, k_\perp)\,, \qquad (3.2)$$

where the diffuse structure factor is given by

$$S_{diff}(\mathbf{k}_\parallel, k_\perp) = \int d^2r\, \Delta C(k_\perp, \mathbf{r})\, e^{i\mathbf{k}_\parallel\cdot\mathbf{r}}\,. \qquad (3.3)$$

The form of the central δ-component is obtained from the identity,

$$\int d^2r\, e^{i\mathbf{k}_\parallel\cdot\mathbf{r}} = (2\pi)^2\, \delta(\mathbf{k}_\parallel). \qquad (3.4)$$

As an example to demonstrate the diffraction line shape from a rough surface, we present a simple numerical simulation of the diffraction from a real surface, the Au:Pd film presented in Fig. 2.2. The diffraction structure factor can be obtained directly from the definition of the kinematic diffraction,

$$S(\mathbf{k}) = A^*(\mathbf{k})\, A(\mathbf{k}) = |\sum \exp[i\, \mathbf{k}_\parallel\cdot\mathbf{r}_j + i\, k_\perp z(\mathbf{r}_j)]|^2,$$

where the data of the surface height, $z(\mathbf{r}_j)$, as functions of $\mathbf{r}_j$, are taken from the digitized STM image.

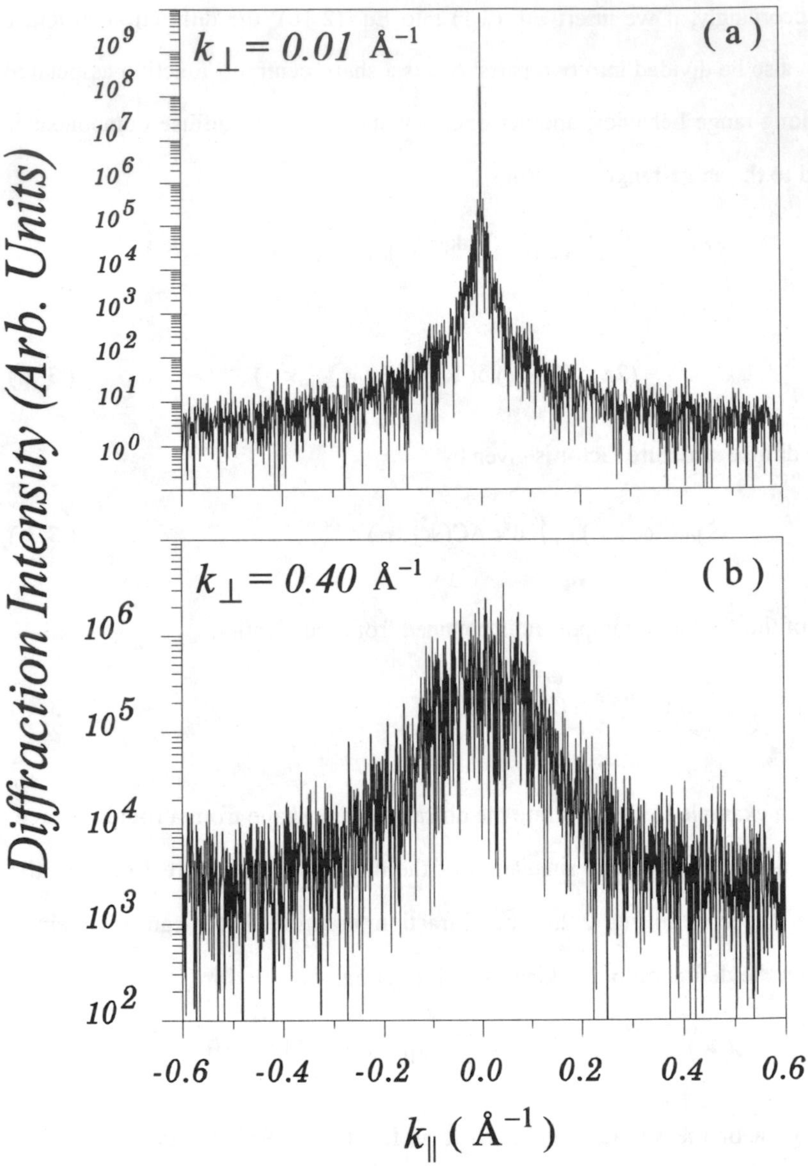

Fig. 3.2 Simulated intensity distribution of the (00) beam "diffracted" from the Au:Pd film shown in Fig. 2.2. The intensity is calculated at: (a) a small $k_\perp$ condition, $k_\perp$=0.01 Å^{-1}; (b) a larger $k_\perp$ condition, $k_\perp$=0.40 Å^{-1}.

Shown in Fig. 3.2(a) is the plot of the line shape of the simulated diffraction structure factor at $k_\perp$=0.01 Å^{-1}, which clearly demonstrates the existence of both the sharp central δ-component and the broad diffuse profile. In contrast, the plot of the line shape shown in Fig. 3.2(b) is a result of the simulation at $k_\perp$= 0.40 Å^{-1}, where the δ–component is too small to be seen.

§III.1.2 *The δ–component*

We may recall in §I.2.2 and §I.3 that when a diffraction wave is scattered from an infinitely large and perfectly flat surface corresponding to $w \to 0$, we can obtain only the sharp δ-function diffraction line shape. In contrast, for the present rough surface which is infinitely large but not perfectly flat as described by a finite interface width w, the diffraction line shape shown in Eq. (3.2) contains both a δ-peak and a diffuse profile. Equation (3.2) thus exhibits the correspondence between the diffraction intensity distribution and the rough surface morphology: a sharp central δ-peak appears when the surface looks flat on the long-range scale, while a broad diffuse line shape shows up if the surface is rough on the short-range scale.

Compared with the δ-function form in the case of a perfectly flat surface, the δ-component in Eq. (3.2) has an extra $k_\perp$ dependent factor, $C_\infty(k_\perp) = C(k_\perp, \mathbf{r}\to\infty)$. In a continuous surface, for example, we have

$$C_c(k_\perp, \mathbf{r}\to\infty) = e^{-(k_\perp)^2 w_c^2}, \tag{3.5}$$

according to Eq. (2.19) with $H_c(\mathbf{r}\to\infty) \equiv 2w_c^2$. On the other hand, in a crystalline surface, if ϕ is not close to the out-of-phase condition, we have a similar relation,

$$C_d(k_\perp, \mathbf{r}\to\infty) \approx e^{-[\phi]^2 w_d^2}, \tag{3.6}$$

as given by Eq. (2.23) with $H_d(\mathbf{r}\rightarrow\infty) \equiv 2\,w_d^2$. The δ-peak intensity shown above is thus proportional to a Debye-Waller like factor which is extremely sensitive to the global surface roughness characterized by the interface width w. The form of the δ-intensity, shown as $C(k_\perp,\mathbf{r}\rightarrow\infty)\delta(\mathbf{k}_\parallel)$, implies that the existence of local defects in a non-divergent rough surface cannot modify the sharp δ-component line shape but can reduce significantly the absolute δ-intensity.

§III.1.3 *The Diffuse component*

In §III.1.1, we have shown that the diffuse structure factor is a measure of the short-range behavior in a rough surface. In order to show the general aspects of the diffuse structure factor, we consider the following expression from Eq. (3.3),

$$S_{\mathrm{diff}}(\,\mathbf{k}_\parallel, k_\perp\,) = \int d^2r\,[e^{-\frac{1}{2}\Omega\,H(\mathbf{r})/w^2} - e^{-\Omega}]\,e^{\,i\mathbf{k}_\parallel\cdot\mathbf{r}}\,, \qquad (3.7)$$

where Ω is defined as

$$\Omega \equiv \begin{cases} (k_\perp w_c)^2 & \text{for a continuous surface}\,, \\ ([\phi]w_d)^2 & \text{for a crystalline surface}\,. \end{cases} \qquad (3.8)$$

Equation (3.7) is obtained by inserting Eq. (2.19) or Eq. (2.23) (the lowest order approximation in the height difference function) into $\Delta C(k_\perp, \mathbf{r})$ in Eq. (3.3). We have also made use of Eqs. (3.5) and (3.6) to define $C(k_\perp, \mathbf{r}\rightarrow\infty)$ in Eq. (3.3). As will be shown later, Ω is a critical quantity to determine the rough surface properties in a diffraction experiment.

Equation (3.7) can be applied to either the continuous surface case with $H(\mathbf{r}) = H_c(\mathbf{r})$ and $w = w_c$, or the crystalline surface case, where $H(\mathbf{r}) = H_d(\mathbf{r})c^2$ and $w = w_d c$ for φ not close to the out-of-phase conditions (see Eq. (2.23)). We must be aware that for the crystalline surface case, Eq. (3.7) does not work at the near out-of-phase condition where $C(k_\perp,\mathbf{r})$ has to be represented by the rigorous expression, Eq. (2.22), instead of its lowest order approximation, Eq. (2.23). For the present purpose, throughout most of parts in this chapter, we shall concentrate on the discussion of Eq. (3.7). But we will give a detailed discussion on the out-of-phase diffraction in a special section (§III.4.2).

If we employ the generic form of the height-height correlation function, Eq. (2.27), the diffuse structure factor, Eq. (3.7), becomes

$$S_{diff}(\mathbf{k}_\parallel, k_\perp) = \int d^2r\, [e^{-\Omega f(r/\xi)} - e^{-\Omega}]\, e^{i\mathbf{k}_\parallel\cdot\mathbf{r}}$$

$$= e^{-\Omega} \int d^2r\, \{e^{\Omega[1-f(r/\xi)]} - 1\}\, e^{i\mathbf{k}_\parallel\cdot\mathbf{r}}. \qquad (3.7')$$

For the function, $P(\mathbf{r}) = P(r) = e^{\Omega[1-f(r/\xi)]} - 1$, which is independent of the polar angle, we have

$$\int d^2r\, P(r)\, e^{i\mathbf{k}_\parallel\cdot\mathbf{r}} = \int_0^\infty P(r)\, rdr \int_0^{2\pi} d\theta\, e^{ik_\parallel r\cos\theta} = 2\pi \int_0^\infty rdr\, P(r)\, J_0(k_\parallel r)$$

$$= 2\pi \int_0^\infty rdr\, \{e^{\Omega[1-f(r/\xi)]} - 1\}\, J_0(k_\parallel r),$$

where $J_0(k_\parallel r)$ is the zeroth order Bessel function given by $J_0(k_\parallel r) = \frac{1}{2\pi}\int_0^{2\pi} d\theta\, e^{ik_\parallel r\cos\theta}$.

By expanding P(r) as a Taylor series,

$$P(r) = e^{\Omega[1-f(r/\xi)]} - 1 = \sum_{m=1}^{\infty} \frac{1}{m!}\{\Omega[1 - f(\tfrac{r}{\xi})]\}^m,$$

we can rewrite the diffuse structure factor Eq. (3.7') as

$$S_{diff}(k_{\|}, k_{\perp}) = 2\pi e^{-\Omega} \sum_{m=1}^{\infty} \frac{1}{m!} \Omega^m \int_0^{\infty} rdr \; [1 - f(\tfrac{r}{\xi})]^m J_0(k_{\|} r)$$

$$= 2\pi \xi^2 e^{-\Omega} \sum_{m=1}^{\infty} \frac{1}{m!} \Omega^m \int_0^{\infty} XdX \, [1 - f(X)]^m J_0(k_{\|} \xi X), \quad (3.9)$$

where the last step is obtained by a linear transformation, $r = \xi X$. Equation (3.9) indicates that the diffuse structure factor is a sum of an infinite number of special functions which have the form, $F(m, Y) = \int_0^{\infty} XdX \, [1 - f(X)]^m J_0(YX)$, where $Y = k_{\|}\xi$. This function depends both on the specific form of the scaling function f(X) and on the lateral correlation length ξ.

In order to understand the meaning of Eq. (3.9), let us study a simple example based on the phenomenological scaling function given in Eq. (2.28). If we insert the phenomenological function,[3.2] $f(X) = 1 - e^{-X^{2\alpha}}$, into Eq. (3.9), we obtain

$$S_{diff}(k_{\|}, k_{\perp}) = 2\pi \xi^2 e^{-\Omega} \sum_{m=1}^{\infty} \frac{1}{m!} \frac{1}{m^{1/\alpha}} \Omega^m F_\alpha(k_{\|} \xi m^{-\frac{1}{2\alpha}}), \quad (3.10)$$

where we have $F(m, Y) = \frac{1}{m^{1/\alpha}} F_\alpha(Y m^{-\frac{1}{2\alpha}})$ with

$$F_\alpha(Y') = \int_0^{\infty} X'dX' \, e^{-X'^{2\alpha}} J_0(X'Y'), \quad (Y' = Y m^{-\frac{1}{2\alpha}}). \quad (3.11)$$

We have replaced X by $X' = m^{\frac{1}{2\alpha}} X$. In Fig. 3.3, we plot the line shape $F_\alpha(Y')$ as a function of Y' at several values of α. Assume that Y'_g satisfies $F_\alpha(Y'_g) = 0.5 \, F_\alpha(0)$, where Y'_g is a constant depending only on α through Eq. (3.11). We can then show that the FWHM of the function $F_\alpha(k_{\|}\xi m^{-\frac{1}{2\alpha}})$ measured in $k_{\|}$ is given by FWHM =

$Y'_g\,\xi^{-1} m^{\frac{1}{2\alpha}}$. This simple example gives a specific diffuse structure factor, Eq. (3.10), which is a sum of an infinite number of the function $F_\alpha(Y')$ with different widths $\propto \xi^{-1} m^{\frac{1}{2\alpha}}$.

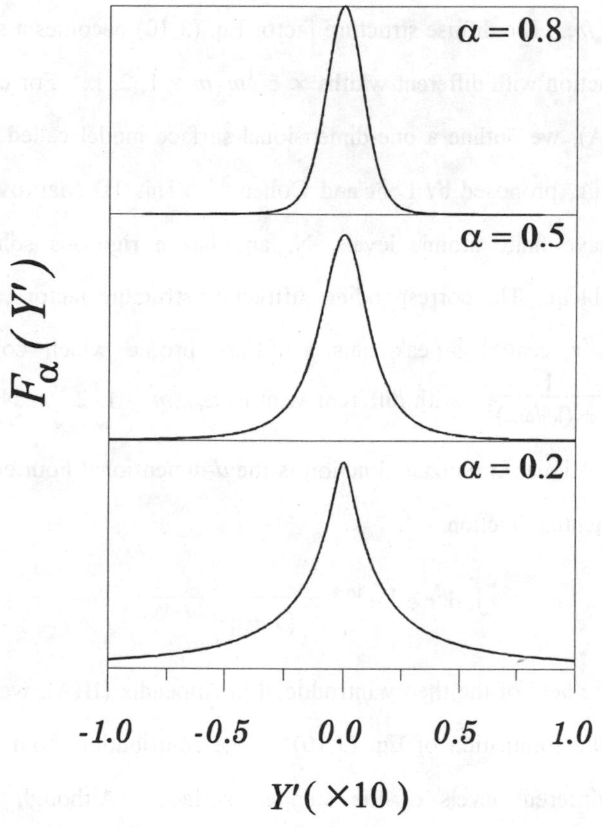

Fig. 3.3 The line shapes of $F_\alpha(Y')$ as a function of Y' for different values of α.

The physical meaning of Eq. (3.10) in this simple example can further be understood for the case of $\alpha = 0.5$. For $\alpha = 0.5$, Eq. (3.11) turns out to be a 2D Lorentzian function,

$$F_{1/2}(Y') \propto \frac{1}{[1 + Y'^2]^{3/2}},$$

where $Y' = k_{||}\xi/m$. The diffuse structure factor Eq. (3.10) becomes a sum of the 2D Lorentzian function with different widths $\propto \xi^{-1}m$, $m = 1, 2, \ldots$ For comparison, in Appendix (IIIA), we outline a one-dimensional surface model called the restricted Markovian chain, proposed by Lent and Cohen [3.3]. This 1D Markovian surface is assumed to have finite atomic levels, N, and has a rigorous solution for the diffraction problem. The corresponding diffraction structure factor can be derived analytically as a central δ-peak plus a diffuse profile which consists of 1D Lorentzians, $\frac{1}{1 + (k_{||}/\omega_m)^2}$, with different widths, ω_m, $m = 1, 2, \ldots, N-1$. We note that the d-dimensional Lorentzian function is the d-dimensional Fourier transform of a similar exponential function, e^{-r},

$$\int d^d\mathbf{r}\, e^{-r}\, e^{\,i\mathbf{q}\cdot\mathbf{r}} \propto \frac{1}{[1 + q^2]^{(d+1)/2}}.$$

With the help of the theory introduced in Appendix (IIIA), we can interpret each term in the summation of Eq. (3.10) as the contribution from the scattering events from different levels of the stepped surface. Although the Gaussian distribution involves many atomic levels, most of the height fluctuations should still be confined within the range of the interface width w which is a finite quantity for this non-diverging interface. This fact is also reflected in Eq. (3.10) where the coefficient, $\frac{1}{m!}\frac{1}{m^{1/\alpha}}\Omega^m$, of the F_α-function would quickly vanish as $m > \Omega$. The number of 2D Lorentzians, M, which gives a significant contribution to the diffuse structure factor, is therefore finite and determined by the condition,

$$M \sim \Omega. \tag{3.12}$$

We can safely neglect all of the small Lorentzian terms with $m >> \Omega$. Equation (3.10) then becomes the sum of a finite number of 2D Lorentzian functions, where the finite number M is limited by the interface width w through Eqs. (3.12) and (3.8). From this point of view, the two-dimensional rough surface with $\alpha = 0.5$ looks like a "2D restricted Markovian surface" analogous to the 1D restricted Markovian chain detailed in Appendix (IIIA).

The conclusions drawn from the simple example shown above can be generalized to the cases described by Eq. (3.9). In conclusion, the diffuse structure factor from a rough surface consists of an infinite number of special functions with different widths. The line shapes of the special functions as well as their widths depend on the roughness parameter α and the correlation length ξ. However, the number M of the special functions, which have a significant contribution to the diffuse structure factor, is finite, and is determined by the interface width w through Eq. (3.12).

§III.2 Diffraction Structure Factor from Rough Surfaces: Asymptotic Behavior

Although the diffraction structure factor shown in Eqs. (3.2) and (3.9) looks complicated, its asymptotic behavior is relatively simple. In this section, we shall discuss two extreme cases, which are $\Omega << 1$ and $\Omega >> 1$, respectively.

§III.2.1 *Diffraction Structure Factor for* $\Omega << 1$

The condition, $\Omega << 1$, is easily realized in a practical diffraction experiment. For a continuous surface, the condition $\Omega = (k_\perp w_c)^2 << 1$ corresponds to small $k_\perp$. For a crystalline surface, $\Omega = ([\phi]w_d)^2 << 1$ can be applied at the near in-phase diffraction conditions.

Under the condition, $\Omega << 1$, only the first term (with respect to Ω) would give a significant contribution to the diffuse structure factor, Eq. (3.9). We can therefore represent the diffraction structure factor as

$$S(\mathbf{k}) \approx (2\pi)^2 e^{-\Omega}\delta(\mathbf{k}_{||}) + 2\pi\,\Omega\,\xi^2 e^{-\Omega}F(m=1, k_{||}\xi), \quad (\Omega << 1), \qquad (3.13)$$

where $F(m{=}1,\ Y) = \int_0^\infty XdX[1- f(X)]J_0(XY)$. If the phenomenological scaling function of Eq. (2.28) is used, we have $F(m{=}1, Y) = F_\alpha(Y)$ as shown in Eq. (3.11), where $Y = Y' = k_{||}\xi$.

The line shape of the diffuse structure factor in Eq. (3.13) has a FWHM inversely proportional to the lateral correlation length,

$$\text{FWHM} = \frac{2Y_g}{\xi}, \qquad (3.14)$$

where the constant Y_g satisfies $F(m{=}1, Y_g) = 0.5F(m{=}1, 0)$. This relationship can be utilized for the determination of the lateral correlation length in a diffraction experiment.

On the other hand, the contribution of the diffuse structure factor is very small as compared to that of the δ–component intensity. The peak intensity of the diffraction structure factor shown in Eq. (3.13) is therefore approximately proportional to the Debye-Waller like factor, $e^{-\Omega}$, where Ω is defined in Eq. (3.8). As an example, we examine the simulated diffraction intensity from the Au:Pd film shown in Fig. 2.2. We plot in Fig. 3.4 the simulated peak intensity vs. $k_\perp$ as open squares. The solid curve in Fig. 3.4 is the fit using the Debye-Waller-like function, $e^{-(k_\perp w)^2}$, where w is an adjustable parameter. In the small $k_\perp$ regime, $0 \le k_\perp < 0.15$ $Å^{-1}$, the quality of fit is very good. However, as shown in Fig 3.4, the deviation between the fitted curve and the "experimental data" becomes significant at the larger $k_\perp$ regime, 0.15 $Å^{-1} < k_\perp < 0.30$ $Å^{-1}$, where the contribution from the diffuse intensity becomes comparable with that from the δ–intensity. Actually, the δ–component will become negligibly small for $k_\perp > 0.30$ $Å^{-1}$, as demonstrated in the plot of the intensity distribution for $k_\perp = 0.40$ $Å^{-1}$ shown in Fig. 3.2(b).

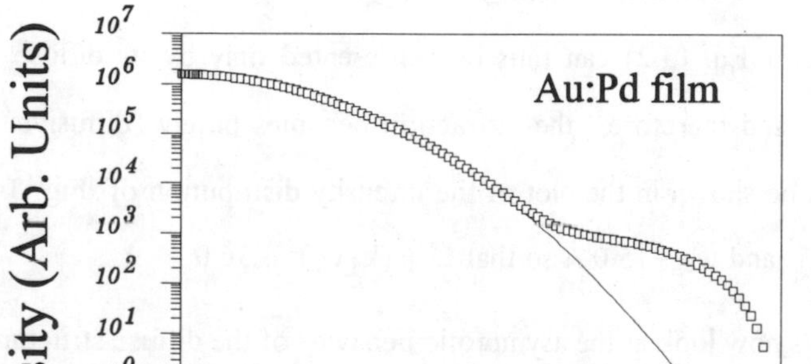

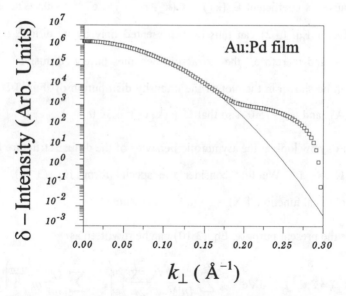

Fig. 3.4 The simulated peak intensity "diffracted" from the Au:Pd film shown in Fig. 2.2 as a function of $k_\perp$ (open squares). The solid curve is the Debye-Waller-like function, $e^{-(k_\perp w)^2}$, where w is an adjustable parameter to give a best fit.

§III.2.2 *Diffraction Structure Factor for* $\Omega >> 1$

$\Omega >> 1$ can be realized under two conditions. One is that the surface is rough with a sufficiently large interface width w. Another condition involves the diffraction condition. In order to have $\Omega = (k_\perp w_c)^2 >> 1$ in a continuous surface, the vertical wavevector $k_\perp$ must be sufficiently large, which is just opposite to the small $k_\perp$ diffraction condition. Accordingly, for a crystalline surface, $\Omega = ([\phi]w_d)^2 >> 1$ can be easily satisfied at the diffraction conditions $|[\phi]| > 0$. For w_d as small as 5, $([\phi]w_d)^2$ is as large as ~ 140 at $[\phi] = 0.75\pi$ and ~ 60 at $[\phi] = 0.5\pi$. However, at this point we still would like to avoid the condition that $[\phi]$ is too close to π in which Eqs. (3.6) and (3.7) do not hold.

Under the condition, $\Omega >> 1$, the δ–component in Eq. (3.2) is negligibly small, because its coefficient $C_\infty(k_\perp) = C(k_\perp, \mathbf{r}\to\infty) \approx e^{-\Omega} \to 0$. The diffraction structure factor Eq. (3.2) can thus be represented only by its diffuse component, $S_{diff}(\mathbf{k}_{||}, k_\perp)$, and therefore, the diffraction becomes purely "diffusive". A typical example can be shown in the plot of the intensity distribution of Fig. 3.2(b), where $k_\perp = 0.40$ Å^{-1} and $w_c \sim 15.0$ Å so that $\Omega = (k_\perp w_c)^2 \sim 36.0 >> 1$.

Let us now look at the asymptotic behavior of the diffuse structure factor Eq. (3.9) for $\Omega >> 1$. We first consider the special form, Eq. (3.10), where the phenomenological function, $f(X) = 1 - e^{-X^{2\alpha}}$, is assumed.

For the present purpose, Eq. (3.10) can be rewritten as

$$S_{diff}(\mathbf{k}_{||}, k_\perp) = 2\pi\xi^2 \int_0^\infty X'dX' e^{-X'^{2\alpha}} \sum_{n=0}^{\infty} \frac{(-1)^n}{(n!)^2} \left(\frac{k_{||}\xi X'}{2}\right)^{2n} \left(e^{-\Omega} \sum_{m=1}^{\infty} \frac{\Omega^m}{m!} m^{-\frac{1+n}{\alpha}} \right), \quad (3.10')$$

where we expand the zeroth order Bessel function,

$$J_0(k_{\|}\xi X') = \sum_{n=0}^{\infty} \frac{(-1)^n}{(n!)^2}\left(\frac{k_{\|}\xi X'}{2}\right)^{2n}. \qquad (3.15)$$

In the case of $\Omega >> 1$, we have a useful asymptotic identity (the rigorous proof is shown in the Appendix (IIIB)),

$$e^{-\Omega}\sum_{m=1}^{\infty}\frac{\Omega^m}{m!}m^{-\frac{1+n}{\alpha}} \approx \Omega^{-\frac{1+n}{\alpha}}, \qquad (\Omega >> 1).$$

With this asymptotic behavior, we can simplify Eq. (3.10') as

$$S_{diff}(\mathbf{k}_{\|}, k_{\perp}) \approx 2\pi\xi^2\int_0^{\infty} X'dX'e^{-X'^{2\alpha}}\sum_{n=0}^{\infty}\frac{(-1)^n}{(n!)^2}\left(\frac{k_{\|}\xi X'}{2}\right)^{2n}\Omega^{-\frac{1+n}{\alpha}}$$

$$= 2\pi\xi^2\Omega^{-\frac{1}{\alpha}}\int_0^{\infty} X'dX'e^{-X'^{2\alpha}}J_0(k_{\|}\xi\Omega^{-\frac{1}{2\alpha}}X').$$

We therefore obtain the diffraction structure factor at $\Omega >> 1$ as a simple form,

$$S_{diff}(\mathbf{k}_{\|}, k_{\perp}) \approx 2\pi\,(\xi\Omega^{-\frac{1}{2\alpha}})^2\,F_{\alpha}(k_{\|}\,\xi\Omega^{-\frac{1}{2\alpha}}), \quad (\Omega >> 1), \qquad (3.16)$$

where the function $F_{\alpha}(Z)$ is defined in Eq. (3.11) with $Z = k_{\|}\,\xi\Omega^{-\frac{1}{2\alpha}}$. Note that the sum of $F_{\alpha}(Y')$ function in Eq. (3.10) has collapsed into a single $F_{\alpha}(Z)$ function with a different argument. Although Eq. (3.16) is obtained from Eq. (3.10) for a special scaling function, $f(X) = 1 - e^{-X^{2\alpha}}$, we can show that this asymptotic structure factor is "universal" for any scaling functional form f(X) defined in Eq. (2.27). The rigorous proof of Eq. (3.16) for an arbitrary f(X) is shown in Appendix (IIIC). The physical reason for this universal asymptotic behavior can be explained as follows.

The diffraction structure factor shown in Eq. (3.16) is related to the parameter, $\xi\Omega^{-\frac{1}{2\alpha}}$, which, according to Eq. (3.8), can be expressed as

$$\xi\Omega^{-\frac{1}{2\alpha}} = \begin{cases} \xi w_c^{-\frac{1}{\alpha}} (k_\perp)^{-\frac{1}{\alpha}} & \text{for a continuous surface ,} \\ \xi w_d^{-\frac{1}{\alpha}} [\phi]^{-\frac{1}{\alpha}} & \text{for a crystalline surface .} \end{cases} \qquad (3.17)$$

As we have pointed out in §II.3.2, $\xi w^{-\frac{1}{\alpha}} = \eta$ is an important parameter characterizing the short-range properties in a rough surface. Equation (3.16) therefore depends only on the short-range parameters, α and η. The fact that the diffraction at $\Omega >> 1$ only detects the short-range behavior in a rough surface can also be understood from Fig. 3.1(c). As shown in Fig. 3.1(c), $C(k_\perp, \mathbf{r})$ at $\Omega >> 1$ is clearly confined to the short-range regime. Analytically, the short-range height difference function from a rough surface can be expressed as

$$C(k_\perp, \mathbf{r}) = e^{-\frac{1}{2}\Omega H(\mathbf{r})/w^2} = e^{-\Omega f(r/\xi)} \approx e^{-\Omega (r/\xi)^{2\alpha}} = e^{-(\Omega/\xi^{2\alpha})\, r^{2\alpha}}, \quad (r << \xi), \qquad (3.18)$$

where we have used the asymptotic form of the scaling function, $f(X) \approx X^{2\alpha}$, for $X << 1$. If the short-range height difference function Eq. (3.18) is inserted into the original diffraction structure factor defined by Eq. (2.16), we can immediately obtain Eq. (3.16). We therefore conclude that the asymptotic diffraction structure factor, Eq. (3.16), is the Fourier transform of the height difference function in the short-range regime given by Eq. (3.18). The universal form of the short-range height difference function thus leads to the universal form of the diffraction structure factor characterized by η and α.

The FWHM of Eq. (3.16) measured in $\mathbf{k}_\parallel$ can be written as

$$\text{FWHM} = \begin{cases} 2Z_g\eta^{-1}(k_\perp)^{\frac{1}{\alpha}} & \text{for a continuous surface}, \\ 2Z_g\eta^{-1}\|[\phi]\|^{\frac{1}{\alpha}} & \text{for a crystalline surface}, \end{cases} \tag{3.19}$$

where Z_g satisfies $F_\alpha(Z_g) = 0.5\, F_\alpha(0)$. In analog to the model shown in Appendix (VIA), η is interpreted as proportional to the average terrace size in the case of crystalline rough surfaces. Also, in this case, as we shall discuss more in §III.4.2, the FWHM oscillates as a function of ϕ between in-phase and out-of-phase conditions.

In conclusion, we can express the diffraction structure factor for $\Omega >> 1$ as

$$S(\mathbf{k}_\|, k_\perp) \sim \begin{cases} (\eta k_\perp^{-\frac{1}{\alpha}})^2\, F_\alpha(k_\|\eta k_\perp^{-\frac{1}{\alpha}}) & \text{for a continuous surface}, \\ (\eta[\phi]^{-\frac{1}{\alpha}})^2\, F_\alpha(k_\|\eta[\phi]^{-\frac{1}{\alpha}}) & \text{for a crystalline surface}. \end{cases} \tag{3.20}$$

§III.3 Measurement of w, ξ And α from Diffraction Experiments

Based on the characteristics of the diffraction structure factor shown in §III.1 and §III.2, we are now able to extract from a diffraction experiment the interface width w, the lateral correlation length ξ, and the roughness exponent α.

§III.3.1 *Measurement of interface width w*

According to Eqs. (3.5) and (3.6), the δ–intensity has a coefficient,

$$C_\infty(k_\perp) = C(k_\perp, \mathbf{r} \to \infty) = e^{-\Omega},$$

where Ω is defined by Eq. (3.8). This relationship provides a method to obtain w through the diffraction line shape measurement. Since the term, $e^{-\Omega}$, decays very quickly with the increase of the vertical wavevector of $k_\perp$, a practical measurement can only be performed for small $k_\perp$ (for a continuous surface) or for the near in-phase condition (for a crystalline surface). There are a number of ways to obtain the interface width w.

(i) Extract w from the measured curve of I_δ vs. $k_\perp$ (or I_δ vs. $[\phi]$), where I_δ is the δ–intensity. According to the relation, $I_\delta \propto e^{-(k_\perp w_c)^2}$, (or $I_\delta \propto e^{-[\phi]^2 w_d^2}$), we can plot the measured curve of $\mathrm{Ln}(I_\delta)$ vs. $(k_\perp)^2$, (or $\mathrm{Ln}(I_\delta)$ vs. $[\phi]^2$). The slope of the plot then gives the interface width w^2.

(ii) Extract w from the measured ratio, R_δ, of the integrated δ–intensity to the total integrated diffraction intensity. In Eq. (3.2), we can derive the following relations.

$$\int d^2k_{\|}\, S_\delta(\mathbf{k}_{\|}, k_\perp) = (2\pi)^2\, e^{-\Omega},$$

where $S_\delta(\mathbf{k}_{\|}, k_\perp) = (2\pi)^2\, e^{-\Omega}\, \delta(\, \mathbf{k}_{\|}\,)$. Also,

$$\int d^2k_{\|}\, S(\,\mathbf{k}\,) = \int d^2k_{\|} \int d^2r\, C(\,k_\perp, \mathbf{r}\,)\, e^{\,i\mathbf{k}_{\|}\cdot\mathbf{r}}$$

$$= \int d^2r\, C(\,k_\perp, \mathbf{r}\,) \int d^2k_{\|}\, e^{\,i\mathbf{k}_{\|}\cdot\mathbf{r}} = (2\pi)^2,$$

where we have used the identities Eq. (3.4) and $C(k_\perp, \mathbf{r} = 0) = 1$. The ratio R_δ of the integrated intensities can then be given by

$$R_\delta = \frac{\int d^2k_{\|}\, S_\delta(\mathbf{k}_{\|}, k_\perp)}{\int d^2k_{\|}\, S(\,\mathbf{k}\,)} = e^{-\Omega},$$

which has a similar relation discussed in (i).

(iii) Extract w from the measured ratio, R_D, of the integrated diffuse intensity to the total integrated diffraction intensity. Since

$$\int d^2k_{\|} S_{diff}(\mathbf{k}_{\|}, k_{\perp}) = \int d^2k_{\|} \int d^2r \, \Delta C(k_{\perp}, \mathbf{r}) \, e^{i\mathbf{k}_{\|}\cdot\mathbf{r}}$$

$$= \int d^2r \, \Delta C(k_{\perp}, \mathbf{r}) \int d^2k_{\|} \, e^{i\mathbf{k}_{\|}\cdot\mathbf{r}} = (2\pi)^2 [1 - e^{-\Omega}],$$

we have

$$R_D = \frac{\int d^2k_{\|} S_{diff}(\mathbf{k}_{\|}, k_{\perp})}{\int d^2k_{\|} S(\mathbf{k})} = 1 - e^{-\Omega}.$$

In the case of $\Omega << 1$, the relationship is simplified as $R_D \approx \Omega$. This simple relationship can be most easily realized in light scattering measurements where $k_{\perp} \sim 10^{-3}$ Å^{-1} $<< w^{-1} \sim 0.1$ Å^{-1} so that $\Omega = (k_{\perp}w)^2 << 1$.

In the case of HRLEED, care must be taken in extracting the interface width w due to the effect of the multiple scattering. If the multiple scattering effect is included, the diffraction intensity can be modified to give approximately[3,4],

$$I(\mathbf{k}) \propto Q(E, \theta) \, S(\mathbf{k}), \qquad (3.21)$$

where $Q(E, \theta)$ is the modified factor resulting from the multiple scattering effect. This factor depends on the primary electron energy, $E = \frac{1}{2m_e} k^2 \propto k_{\|}^2 + k_{\perp}^2$, and the incident angle, θ, of the electron beam with respect to the normal direction. In the case of a HRLEED experiment, $\theta \approx 0$, we therefore have, $E \propto k_{\perp}^2$. Since the multiple scattering factor $Q(E, \theta)$ is only sensitive to the vertical wavevector, $k_\perp$, in a HRLEED experiment, one can express Eq. (3.21) as

$$I(\mathbf{k}) \propto Q(k_\perp)\, S(\mathbf{k})\,. \tag{3.21'}$$

According to Eq. (3.21'), the simple method to extract w, as shown in (i), is not valid due to the modification of $I(\mathbf{k})$ by the multiple scattering factor, $Q(k_\perp)$. In order to get rid of $Q(k_\perp)$, one needs to do a calibration first with respect to the flat surface. However, the methods shown in (ii) and (iii) are still valid because the calculation of the intensity ratio has automatically canceled the factor $Q(k_\perp)$.

§III.3.2 *Measurement of lateral correlation length* ξ

As shown in Eqs. (3.13) and (3.14), the FWHM of the diffuse structure factor at $\Omega << 1$ is inversely proportional to ξ. The measurement can be realized at small $k_\perp$ diffraction condition (for a continuous surface) or near in-phase condition (for a crystalline surface).

§III.3.3 *Measurement of roughness parameter* α

Experimentally, the exponent α can be measured from the diffraction line shape under the condition $\Omega >> 1$ where the diffraction structure factor is determined only by the short-range parameter, α and η. According to Eqs. (3.16) and (3.17), we have

$$S(0, k_\perp) \propto \begin{cases} (k_\perp)^{-\frac{2}{\alpha}} & \text{for a continuous surface}\,, \\ \|[\phi]\|^{-\frac{2}{\alpha}} & \text{for a crystalline surface}\,. \end{cases} \tag{3.22}$$

Therefore, from the log-log plot of the peak intensity vs. $k_\perp$ or (peak intensity vs. $|[\phi]|$), we can obtain a linear relation in which the slope is given by $-2/\alpha$. This method to obtain α through the power-law shape of the peak intensity can be employed in the X-ray diffraction technique. For a HRLEED experiment, the multiple scattering effect will modify the peak intensity as a function of $k_\perp$, as shown in Eq. (3.21'). We therefore have to be cautious.

On the other hand, from Eq. (3.19), one has

$$\text{FWHM} \propto \begin{cases} (k_\perp)^{\frac{1}{\alpha}} & \text{for a continuous surface ,} \\ |[\phi]|^{\frac{1}{\alpha}} & \text{for a crystalline surface .} \end{cases} \qquad (3.19')$$

This indicates that from the log-log plot of FWHM vs. $k_\perp$ (or FWHM vs. $[\phi]$), we can obtain a linear relation in which the slope is given by $1/\alpha$. The method to obtain α through the FWHM power-law shape can be employed in HRLEED diffraction technique. The multiple scattering effect should not modify the measured FWHM of the line shape because the factor $Q(k_\perp)$ shown in Eq. (3.21') is not sensitive to $k_{\|}$, the parallel wavevector.

We must emphasize that for a crystalline surface, the power-law relationships shown above are not valid in the vicinity of the out-of-phase condition, $|[\phi]| \sim \pi$. As will be shown in next section, the simple power law relationships can be significantly distorted by the "discrete lattice effect" at out-of-phase diffraction conditions.

§III.4 Discrete Lattice Effect on the Diffraction from Rough Crystalline Surfaces

The discrete lattice nature in the crystalline surface can modify both the δ–intensity and the diffuse angular profile at the out-of-phase diffraction conditions.

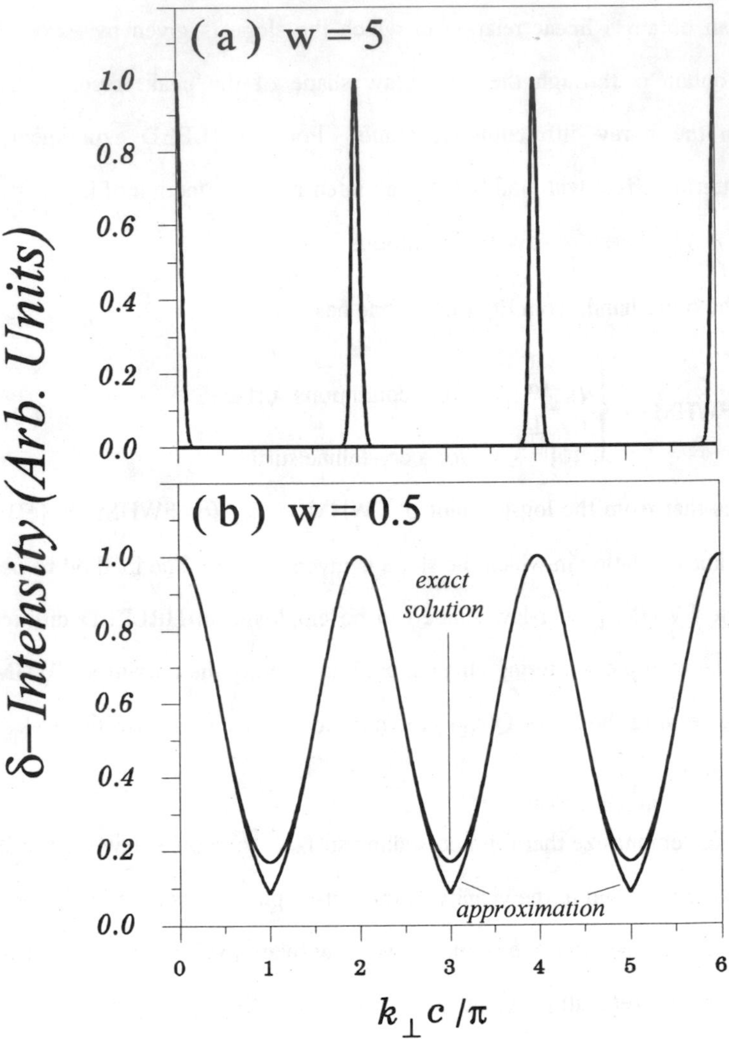

Fig. 3.5 Plots of δ-intensity vs. $k_\perp$ from a crystalline surface for: (a) w = 5 and (b) w = 0.5 (in units of *c*). The curves are calculated, respectively, using Eq. (3.23) (exact solution) and Eq. (3.6) (approximation). Note that in Fig. (a) both calculations give the same curve.

§III.4.1 *The δ–intensity*

Due to its discrete periodic nature, the δ-peak intensity for a crystalline surface, can oscillate between the in-phase and the out-of-phase diffraction conditions. Figures 3.5(a) and 3.5(b) show the comparison between the curves obtained from the rigorous expression for the δ-intensity component,

$$C_\infty(k_\perp) = C_d(k_\perp, \mathbf{r} \to \infty) \approx \frac{\sum_{n=-\infty}^{+\infty} e^{-w_d^2 (\phi - 2\pi n)^2}}{\sum_{n=-\infty}^{+\infty} e^{-w_d^2 (2\pi n)^2}}, \qquad (3.23)$$

and its lowest order approximation, Eq. (3.6), at two different interface widths, respectively. The phase varies from the in-phase to the out-of-phase conditions.

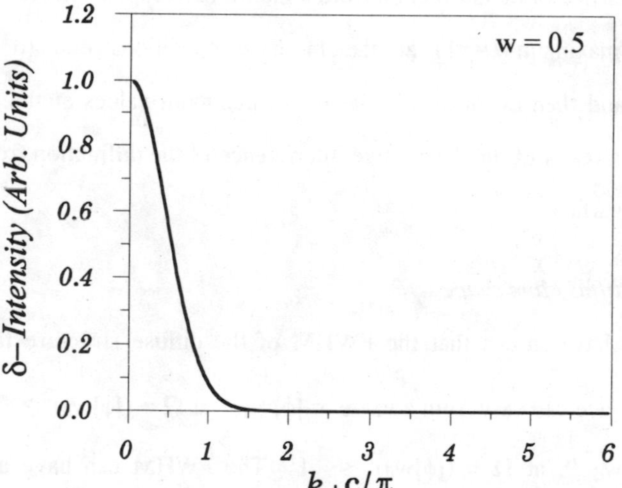

Fig. 3.6 A plot of δ-intensity vs. $k_\perp$ for w = 0.5 from a continuous surface based on Eq. (3.5). In order to compare with Fig. 3.5, we use the same vertical units, *c* for w and π/c for $k_\perp$, respectively.

In contrast to the case for a crystalline surface, the δ-peak intensity from a continuous surface decays monotonically with the increase of $k_\perp$, as given by Eq. (3.5). In Fig. 3.6, for comparison, we plot the nonoscillatory δ-intensity based on Eq. (3.5), corresponding to the continuous surface model.

As shown in Fig. 3.5(b), the difference between Eq. (3.23) and Eq. (3.6) is significant at the near out-of-phase conditions when w_d is small. However, such difference can be ignored for a realistically large interface width, because as $w_d > 1$, $C_d(k_\perp, \mathbf{r} \to \infty)$ vanish quickly at $[\phi] \sim \pi$, as shown in Fig. 3.5(a). The vanishing of the δ-intensity indicates that the diffraction line shape is purely diffusive, which is consistent with the asymptotic behavior discussed in §III.2.2. In contrast to the case of a continuous surface shown in Fig. 3.6, Figs. 3.5(a) and 3.5(b) exhibit an oscillatory behavior of the diffraction from a multilevel stepped surface: $C_d(k_\perp,\mathbf{r}\to\infty)$ reaches the maximum (= 1) at the in-phase conditions due to constructive interference, and then exponentially decays to minimum values at the out-of-phase conditions as a result of the destructive interference of the diffraction from the multi-level stepped surface.

§III.4.2 *The diffuse line shape*

In §III.2, we have shown that the FWHM of the diffuse structure factor from a crystalline surface can vary from a value $\propto [\phi]^{\frac{1}{\alpha}}\eta^{-1}$ at $\Omega = ([\phi]w_d)^2 >> 1$ to a value $\propto \xi^{-1} = \eta^{-1}w_d^{-\frac{1}{\alpha}}$, at $\Omega = ([\phi]w_d)^2 << 1$. The FWHM can have an oscillatory behavior with respect to the phase ϕ. Such an oscillation is a result of the discrete lattice effect. It however does not occurs in a continuous surface, where the FWHM increases monotonically with $k_\perp$ at $\Omega = (k_\perp w_c)^2 >> 1$, as given by Eq. (3.19).

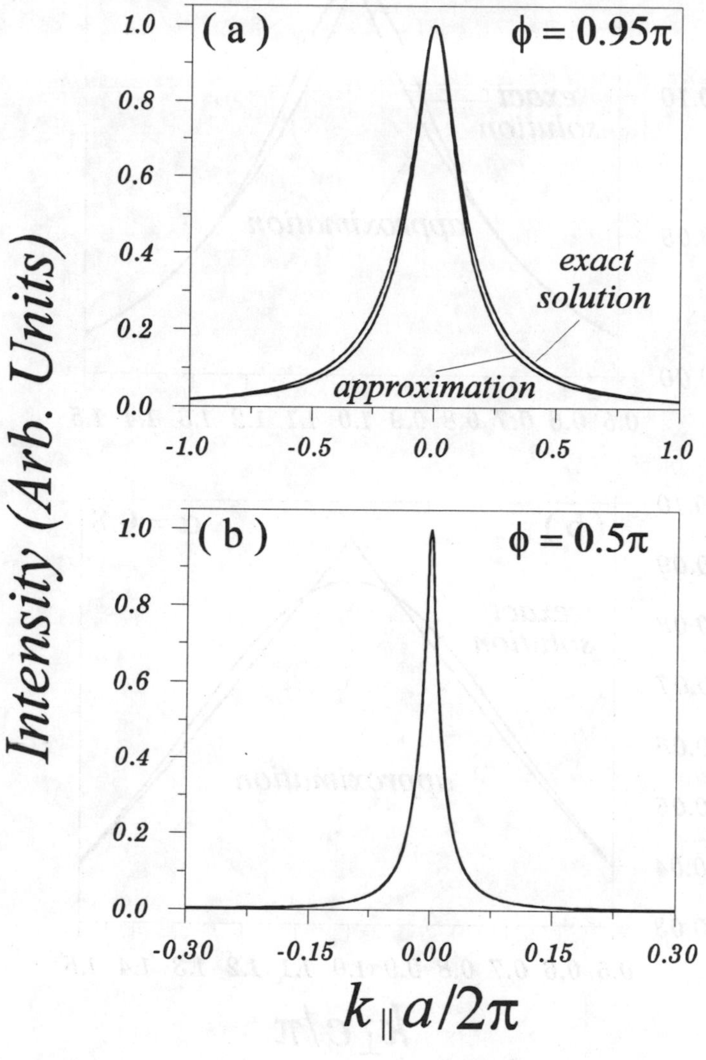

Fig. 3.7 Normalized line shapes of the diffraction structure factor using rigorous and approximate calculations: (a) near out-of-phase condition, ϕ=0.95π; (b) ϕ=0.5π. The difference between the two calculations is more significant in (a).

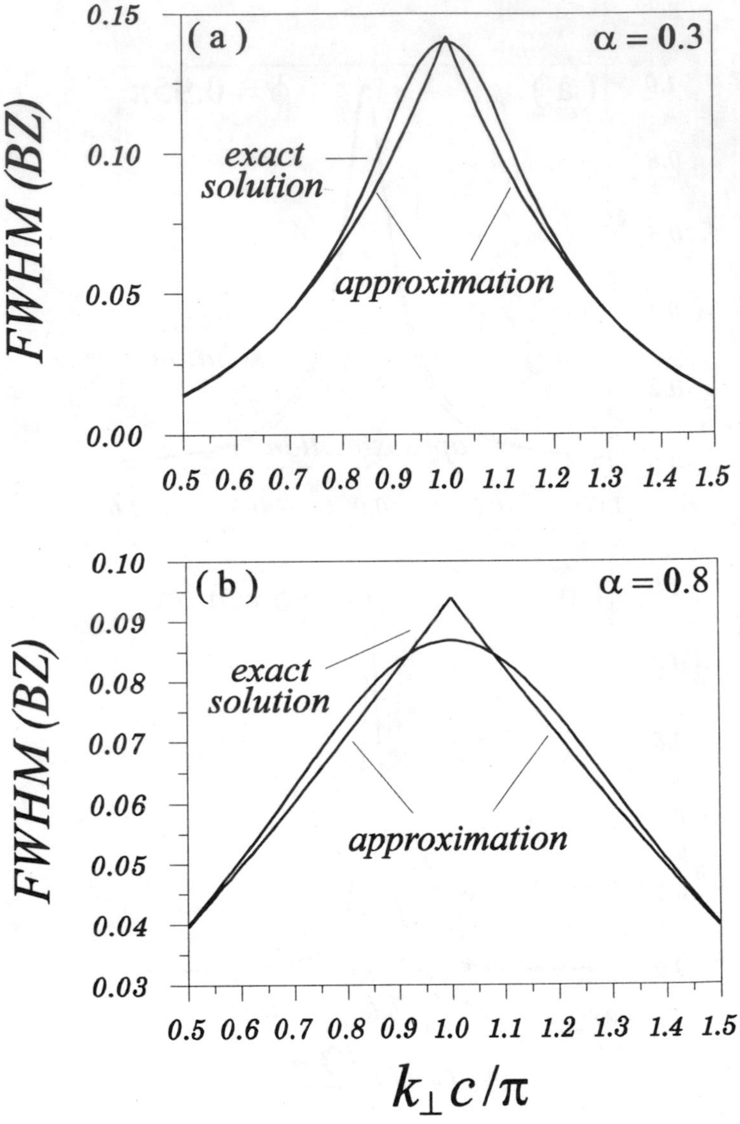

Fig. 3.8 Comparison of the plots of FWHM vs. $k_\perp$ for rigorous and approximate calculations: (a) $\alpha = 0.3$; (b) $\alpha = 0.8$. The FWHM is measured in units of the first Brillouin zone, BZ = $2\pi/a$.

So far, the calculation of the diffuse structure factor from a crystalline surface, Eqs. (3.7) and (3.9), is based on the lowest order approximation, Eq. (2.23), which however, does not apply in the vicinity of the out-of-phase conditions. The rigorous result, which does not have an analytical form, can be calculated numerically by combining Eq. (3.3) with Eq. (2.22). Figures 3.7 are plots of the diffraction structure factor for both the rigorous calculation and the lowest order approximation, Eq. (3.7), respectively at $[\phi] \sim 0.95\pi$ and at $[\phi] \sim 0.5\pi$, where $\alpha = 0.3$, $w = 10.0$ (in units of c) and $\eta = 10.0$ (in units of a). We have used the phenomenological scaling function, $f(X) = 1 - e^{-X^{2\alpha}}$.

As shown in Fig. 3.7, the deviation of the approximation from the rigorous calculation is significant at the near out-of-phase condition ($[\phi] \sim 0.95\pi$, Fig. 3.7(a)), but is negligibly small at $[\phi] \sim 0.5\pi$ (Fig. 3.7(b)). The modification due to the "discrete lattice effect" can also be seen in the plot of FWHM vs. $k_\perp$. In Fig. 3.8, we plot the FWHM of the diffuse line shape as a function of the phase $\phi = k_\perp c$. The two solid curves shown in Fig. 3.8 are respectively the plot based on the rigorous calculation and the plot based on the power-law relation shown in Eq. (3.19). Recall that the power-law relationship of Eq. (3.19) is obtained from the lowest order approximation given by Eq. (3.7) with the asymptotic condition $\Omega >> 1$. The cusp shape at $k_\perp \sim \pi/c$ or $[\phi] \sim \pi$, which is predicted by Eq. (3.19), actually becomes rounded according to the rigorous calculation.

One can have $\Omega = ([\phi]w_d)^2 \sim \pi^2 w_d^2 >> 1$ at the near out-of-phase condition if $w_d > 1$. This means that, $C_d(k_\perp, \mathbf{r} \to \infty) \to 0$, according to Eq. (3.23), i.e., the diffraction structure factor becomes purely diffusive. The diffuse structure factor at the near out-of-phase condition should still be short-range dependent because it is only determined by the short-range height-height correlation given by Eq. (2.25'). Combining the original diffraction structure factor defined by Eq. (2.16) with Eqs. (2.22) and (2.25'), one can easily prove that at the near out-of-phase condition, the diffraction structure factor has a form,

$$S(\mathbf{k}_{\|}, k_{\perp}) \approx S_{diff}(\mathbf{k}_{\|}, k_{\perp}) = H(k_{\|}\eta, \phi), \qquad (\Omega >> 1),$$

where $H(k_{\|}\eta, \phi)$ is periodic in ϕ, $H(k_{\|}\eta, \phi) = H(k_{\|}\eta, \phi + 2\pi)$. Thus, the parameter η can still be interpreted as the average terrace size because FWHM ~ $1/\eta$.

It is found that the shape of FWHM vs. $[\phi]$ plot depends only on α and the magnitude of the FWHM is determined by the value of η. To show this important feature, in Fig. 3.9(a) we plot the calculated curves of the FWHM against $[\phi]$ at various values of η, where the values of α and w_d are fixed. These curves, which have different amplitudes for different η ($= \xi w^{-\frac{1}{\alpha}}$), are based on the rigorous formula by combining Eq. (3.3) with Eqs. (2.22) and (2.27), where the scaling function, $f(X) = 1 - e^{-X^{2\alpha}}$, is used with fixed α (= 0.8) and w (= 1.1). All the curves are then normalized to one at the out-of-phase condition, $[\phi] = \pi$, as shown in Fig. 3.9(b). The normalized plots of FWHM vs. $[\phi]$ exhibit negligibly small differences for different values of η and it therefore implies that the relative variation of the FWHM vs. $[\phi]$ is not sensitive to η but is quite sensitive to α.

We have mentioned before that the methods of extracting α at the near out-of-phase condition from the log-log plots may not be valid in a realistic situation because the "discrete lattice effect" always modifies the simple power law relations described by Eqs. (3.22) and (3.19). As shown in Fig. 3.8(b), the modification due to the "discrete lattice effect" leads to the complete failure of the power law description for $\alpha = 0.8$ at the near out-of-phase condition. (But at α=0.3, as shown in Fig. 3.8(a), in a phase regime which is not close to the out-of-phase, the power law relation can be preserved.) Therefore, at the near out-of-phase diffraction condition, a more reliable method to extract α should be based on the rigorous calculation. One can employ a numerical fitting method to analyze the experimental data. The rigorous relation of FWHM vs. $k_\perp$ for several different values of α are plotted in Fig. 3.10, which can be used to extract the value of α.

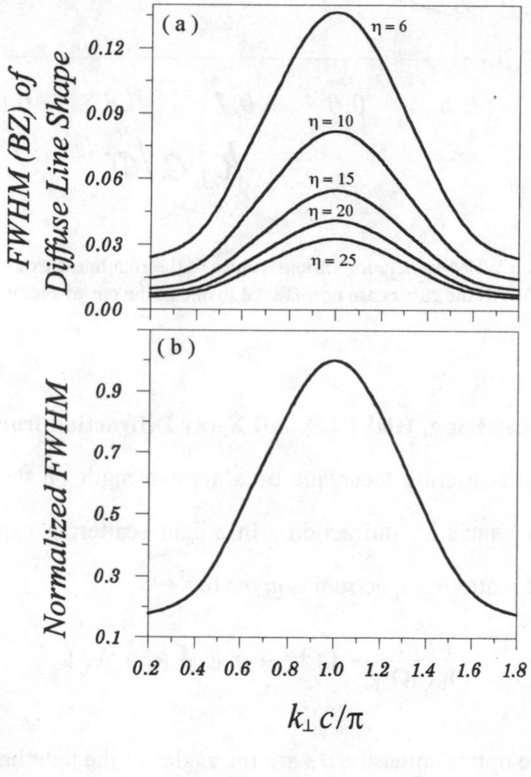

Fig. 3.9 (a) Calculated FWHM vs. $k_\perp$ for various values of η, where the values of α and w_d are fixed; (b) All the curves shown in (a) are normalized to one at the out-of-phase condition, $k_\perp c = \pi$.

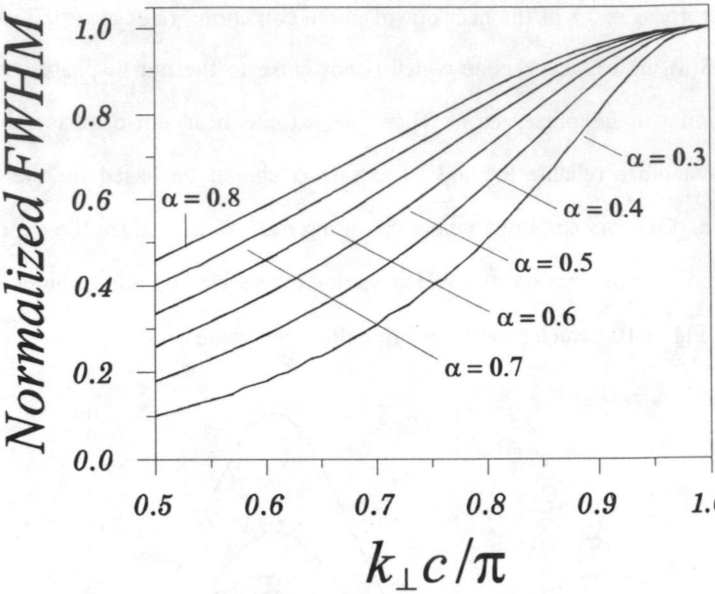

Fig. 3.10 FWHM vs. $k_\perp$, for various values of the roughness exponents, α, at fixed η =10 and w = 5. All the curves are normalized to one at the out-of-phase condition, $k_\perp c = \pi$.

§III.5 Light Scattering, HRLEED and X-ray Diffraction from Rough Surfaces

The visible light scattering technique uses a wavelength on the order of ~ 10^3 Å corresponding to small $k_\perp$ diffraction. In a light scattering experiment, the angle-resolved optical scattering spectrum is given by [3.5]

$$\frac{1}{I_i}\left(\frac{dI}{d\Omega}\right)_s = 4\,k^4 \cos\theta_i \cos^2\theta_s\, R\, W(\,\mathbf{k}_{\|}\,),$$

where I's are the optical intensity, θ's are the angles of the light beams with respect to the surface normal direction and Ω is the solid angle of the illuminated surface area. The subscripts, i and s, denote the incident and scattering directions, respectively. R is a quantity related to the optical reflectivity of the surface. In the angle-resolved optical scattering spectrum, the information of the surface roughness is contained in the optical power spectral density, $W(\mathbf{k}_{\|})$, defined as

$$W(\mathbf{k}_{\|}) = \frac{1}{A} < \left| \frac{1}{2\pi} \int d^2\rho \, e^{i\mathbf{k}_{\|}\cdot\rho} z(\rho) \right|^2 >$$

$$= \frac{1}{A} \frac{1}{(2\pi)^2} \int d^2\rho \int d^2\rho' e^{i\mathbf{k}_{\|}\cdot(\rho - \rho')} < z(\rho) \, z(\rho') >$$

$$\propto \int d^2r \, e^{i\mathbf{k}_{\|}\cdot\mathbf{r}} < z(\mathbf{r}) \, z(0) > ,$$

where A is the area illuminated by the incident light beam and the average height is chosen to be $< z(\mathbf{r}) > = 0$. For the non-divergent rough surface, we have the relation,

$$< z(\mathbf{r}) \, z(0) > = \frac{1}{2} < [z(\mathbf{r})]^2 > + \frac{1}{2} < [z(0)]^2 > - \frac{1}{2} <[z(\mathbf{r}) - z(0)]^2>$$

$$= w^2 - \frac{1}{2} <[z(\mathbf{r}) - z(0)]^2> = w^2[1 - f(\frac{r}{\xi})],$$

according to Eq. (2.27). The optical power spectral density turns out to be

$$W(\mathbf{k}_{\|}) \propto w^2 \, \xi^2 \int_0^\infty XdX[1 - f(X)] \, J_0(k_{\|}\xi X) , \qquad (3.24)$$

which is proportional to the near in-phase diffuse structure factor shown in Eq. (3.13). Equation (3.24) provides the expression of the diffuse line shape in a light scattering experiment. The corresponding central δ–component is expected to be similar to the case discussed in §III.1.

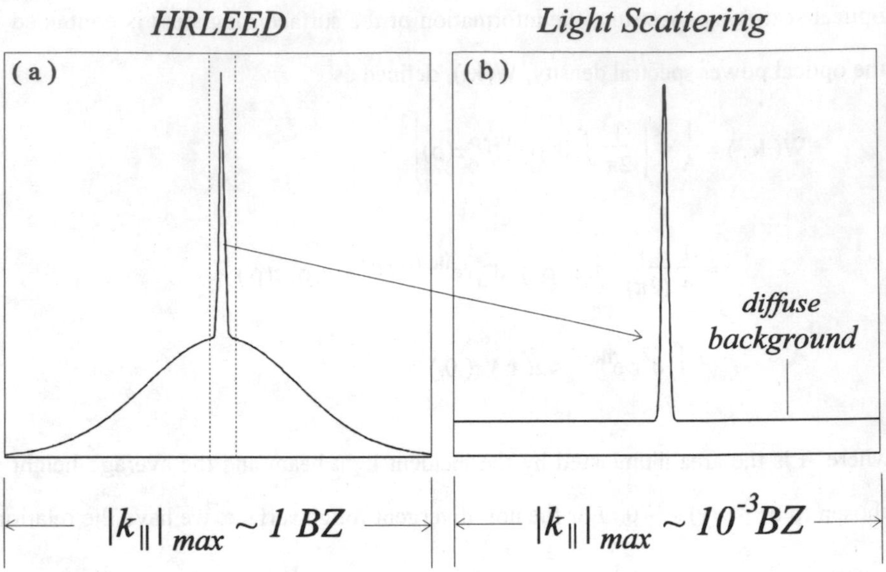

Fig. 3.11 Comparison of diffraction intensity distributions between: (a) HRLEED and (b) light scattering, where the lateral correlation length for a rough surface is assumed to be $\xi << 10^3$ Å.

The comparison between light scattering and HRLEED techniques has been discussed very well by Pietsch *et al.* [3.6] in an instructive way shown in Fig. 3.11 where the lateral correlation length for a rough surface is assumed to be $\xi << 10^3$ Å. To extend the comparison to other ranges of ξ, we plot in Fig. 3.12 the line shapes corresponding to the case of $\xi > 10^3$ Å.

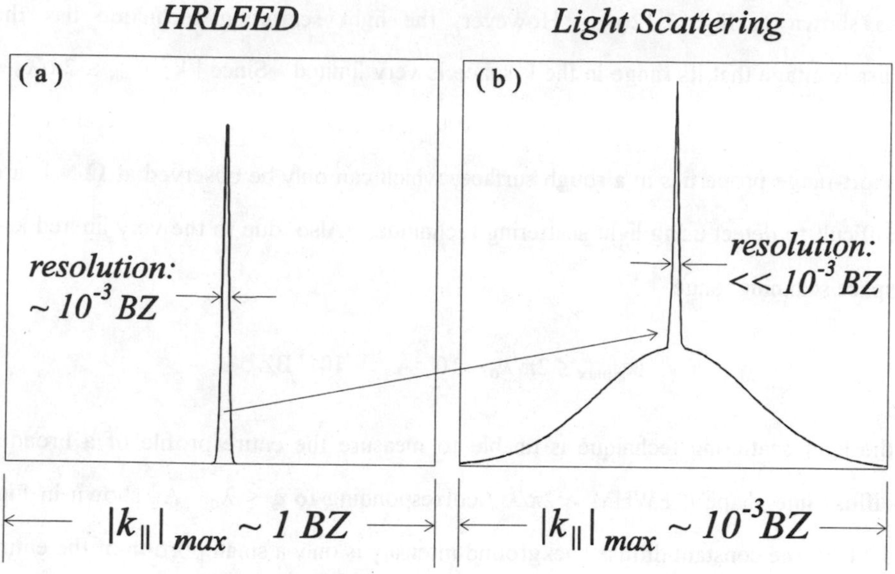

Fig. 3.12 Comparison of diffraction intensity distributions between (a) HRLEED and (b) light scattering in the case of $\xi > 10^3$ Å.

The diffuse light scattering study is more suitable for the rough surface which has a lateral correlation length ξ comparable to the optical wavelength of λ_o, i.e., $\xi \sim \lambda_o \sim 10^3$ Å. As shown in Fig. 3.12(b), the light scattering technique has a very narrow $k_{||}$–space resolution,

$$|\Delta k_{||}|_{resolution} << 2\pi/\lambda_o \sim 10^{-3}\ \text{Å}^{-1} \sim 10^{-3}\ \text{BZ}\,,$$

where BZ = Brillouin zone. One can obtain the detailed fine structures of the diffuse line shape with FWHM $\sim 1/\xi \sim 10^{-3}$ Å^{-1} as $\xi \sim \lambda_o$, according to Eq. (3.24). Neither x-ray nor HRLEED diffraction technique can easily measure such a narrow diffuse profile even with their ultimate $k_{||}$–space resolution, $|\Delta k_{||}|_{resolution} \sim 10^{-3}$ - 10^{-4} Å^{-1}, as shown in Fig. 3.12(a). However, the light scattering technique has the disadvantage that its range in the k-space is very limited. Since $|k_\perp|_{max} \leq 2\pi/\lambda_o \sim 10^{-3}$ Å^{-1}, we have $k_\perp w < 10^{-1}$ for w < 100 Å, i.e., $\Omega = (k_\perp w)^2 << 1$. Therefore, the short-range properties in a rough surface, which can only be observed at $\Omega > 1$, are difficult to detect using light scattering technique. Also, due to the very limited $k_{||}$–space scanning range,

$$|k_{||}|_{max} \leq 2\pi/\lambda_o \sim 10^{-3}\text{Å}^{-1} \sim 10^{-3}\ \text{BZ}\ ,$$

the light scattering technique is unable to measure the entire profile of a broader diffuse line shape if FWHM > $2\pi/\lambda_o$, corresponding to $\xi < \lambda_o$. As shown in Fig. 3.11(b), the constant diffuse background intensity is only a small portion of the entire diffuse line shape that can be measured by the HRLEED as plotted in 3.11(a). Therefore, for a rough surface with its lateral correlation length shorter than the optical wavelength, $\xi < \lambda_o$, the simple evaluation methods shown in §III.3 are in general not valid for a diffuse light scattering experiment. Of course, one can still evaluate the peak intensity of the diffuse line shape, which is proportional to $w^2\xi^2$ according to Eq. (3.24), but the information is more involved since w and ξ cannot be easily separated. For situations like these, it is more appropriate to use X-ray or HRLEED technique which provide a much larger $k_{||}$–space scanning range, $|k_{||}|_{max} \sim 1$ Å$^{-1} \sim 1$ BZ , as shown in Fig. 3.11(a).

REVIEW AND SUMMARY

Diffraction structure factor for rough surfaces with non-divergent interface width

The diffraction structure factor can be divided into a sharp central δ-component and a broad diffuse component,

$$S(\mathbf{k}) = (2\pi)^2 C(k_\perp, \mathbf{r}\to\infty)\delta(\mathbf{k}_\parallel) + S_{diff}(\mathbf{k}_\parallel, k_\perp), \quad (3.2)$$

where the diffuse structure factor is given by

$$S_{diff}(\mathbf{k}_\parallel, k_\perp) = \int d^2r[C(k_\perp,\mathbf{r}) - C(k_\perp,\mathbf{r}\to\infty)]e^{i\mathbf{k}_\parallel\cdot\mathbf{r}}. \quad (3.3)$$

The existence of a δ-component reflects the long-range behavior of the surface having a non-divergent interface width. The diffuse profile is associated with the short-range roughness in the surface.

(1) The δ–component (I_δ)

The I_δ is proportional to the Debye-Waller like factor,

$$C(k_\perp,\mathbf{r}\to\infty) = \begin{cases} e^{-(k_\perp)^2 w_c^2} & \text{for continuous surfaces, (3.5} \\ e^{-[\phi]^2 w_d^2} & \text{for crystalline surfaces, (3.6} \end{cases}$$

where w_c is the interface width for a continuous surface and w_d (in units of *c*) is for a crystalline surface. For the crystalline surface, we have used the lowest order approximation given by Eq. (2.23).

(2) The diffuse component
Away from out-of-phase conditions, the diffuse profile is a sum of an infinite number of special functions,

$$S_{diff}(\mathbf{k}_\parallel, k_\perp) = 2\pi\xi^2 e^{-\Omega}\sum_{m=1}^{\infty}\frac{1}{m!}\Omega^m F(m, k_\parallel\xi), \quad (3.9)$$

where $F(m, Y) = \int_0^\infty XdX\,[1-f(X)]^m J_0(YX)$ and f(X) is the scaling function defined in Eq. (2.27). Ω is an important quantity defined as

$$\Omega = \begin{cases} (k_\perp w_c)^2 & \text{for continuous surfaces,} \\ ([\phi] w_d)^2 & \text{for crystalline surfaces.} \end{cases}$$

Again, for crystalline surfaces, the above conclusion is obtained using the lowest order approximation, Eq. (2.23).

Asymptotic expression of the diffraction structure factor

Based on the results given by Eqs. (3.5), (3.6) and (3.9), (remember that for crystalline surfaces, these are the lowest order approximations), one can further derive asymptotic expressions.

(1) $\Omega << 1$

The diffraction line shape contains a δ-component and a small single diffuse component for $\Omega << 1$,

$$S(\mathbf{k}) \approx (2\pi)^2 e^{-\Omega} \delta(\mathbf{k}_{\|}) + 2\pi\Omega\xi^2 e^{-\Omega} F(m=1, k_{\|}\xi). \quad (3.13)$$

The asymptotic condition, $\Omega << 1$, can be realized at the small $k_\perp$ condition, $k_\perp \sim 0$, for a continuous surface or at the near in-phase condition, $[\phi] \sim 0$, for a crystalline surface.

(2) $\Omega >> 1$

The diffraction line shape is purely diffusive for $\Omega >> 1$, i.e., contains only a broad diffuse profile,

$$S(\mathbf{k}) \sim$$

$$\sim \begin{cases} (\eta k_\perp^{-\frac{1}{\alpha}})^2 F_\alpha(k_{\|} \eta k_\perp^{-\frac{1}{\alpha}}) & \text{(continuous surfaces),} \\ (\eta [\phi]^{-\frac{1}{\alpha}})^2 F_\alpha(k_{\|} \eta [\phi]^{-\frac{1}{\alpha}}) & \text{(crystalline surfaces),} \end{cases} \quad (3.20)$$

where $F_\alpha(Y) = \int_0^\infty X dX e^{-X^{2\alpha}} J_0(XY)$ and $\eta = \xi w^{-\frac{1}{\alpha}}$ ($w = w_c$ or $w = w_d$). For crystalline surfaces, η is proportional to the average terrace size. The asymptotic condition, $\Omega >> 1$, can be realized at the large $k_\perp$ condition, $k_\perp >> 1/w_c$, for a continuous surface or at the diffraction condition away from the "in-phase", $|[\phi]| >> 1/w_d$, for a crystalline surface.

Measurements of the roughness parameters from diffraction experiments

(1) The measurement of w

The w can be measured at $\Omega << 1$, i.e., at the small $k_\perp$ condition, $k_\perp \sim 0$, for a continuous surface or at the near in-phase condition, $[\phi] \sim 0$, for a crystalline surface. There are several ways to do this.

(i) Extract w from the measured curve of δ–intensity (I_δ) vs. $k_\perp$ (or I_δ vs. $[\phi]$), according to the relation,

$$I_\delta \propto e^{-(k_\perp w_c)^2}, \text{(or } I_\delta \propto e^{-[\phi]^2 w_d^2}\text{)}.$$

(ii) Extract w from the measured ratio, R_δ, of the integrated δ–intensity to the total integrated diffraction intensity, according to the relation,

$$R_\delta = \frac{\int d^2k_{\|}\, I_\delta(\mathbf{k})}{\int d^2k_{\|}\, I(\mathbf{k})} = e^{-\Omega} \propto e^{-(k_\perp w_c)^2}, \text{ (or } \propto e^{-[\phi]^2 w_d^2}\text{)}.$$

(iii) Extract w from the measured ratio, R_D, of the integrated diffuse intensity to the total integrated diffraction intensity, according to the relation,

$$R_D = \frac{\int d^2k_{\|}\, I_{diff}(\mathbf{k})}{\int d^2k_{\|}\, I(\mathbf{k})} = 1 - e^{-\Omega}.$$

(2) The measurement of ξ

The ξ can be extracted from the FWHM of the *diffuse structure factor* at $\Omega << 1$ according to the relation, FWHM $\propto 1/\xi$, from Eq. (3.13).

(3) The measurement of α

The α can be measured at $\Omega >> 1$, i.e., at the large $k_\perp$ condition, $k_\perp >> 1/w_c$, for a continuous surface or at the diffraction condition away from the "in-phase", $|[\phi]| >> 1/w_d$, for a crystalline surface. There are two ways to do this.

(i) Extract α from the measured diffraction peak intensity $I(k_{\|}=0, k_\perp)$ vs. $k_\perp$, according to the power-law relation,

$$I(0, k_\perp) \propto \begin{cases} (k_\perp)^{-2/\alpha} & \text{for continuous surfaces,} \\ \|[\phi]\|^{-2/\alpha} & \text{for crystalline surfaces.} \end{cases} \quad (3.22)$$

(ii) Extract α from the measured FWHM vs. $k_\perp$, according to the power-law relation,

$$\text{FWHM} \propto \begin{cases} (k_\perp)^{1/\alpha} & \text{for continuous surfaces,} \\ \|[\phi]\|^{1/\alpha} & \text{for crystalline surfaces.} \end{cases} \quad (3.19)$$

Note that above methods do not apply at near out-of-phase conditions for crystalline surfaces due to the discrete lattice effect.

Discrete lattice effect on the diffraction structure factor

(1) δ–intensity
At the near out-of-phase condition, one should use

$$I_\delta(k_\perp) \propto \frac{\sum_{n=-\infty}^{+\infty} e^{-w_d^2 (\phi - 2\pi n)^2}}{\sum_{n=-\infty}^{+\infty} e^{-w_d^2 (2\pi n)^2}}, \quad (3.23)$$

for the δ–intensity instead of the lowest order approximation given by Eq. (3.6). This formula is also applicable to all values of $\phi = k_\perp c$. Based on this equation, $I_\delta(k_\perp)$ oscillates as a function of ϕ.

(2) Diffuse intensity
In the vicinity of out-of-phase conditions, the lowest order approximation shown before (such as Eq. (3.9)) does not hold anymore for the diffuse intensity. The diffraction line shape deviates significantly from the power-law behavior shown in Eqs. (3.19) and (3.22). The rigorous line shape for these diffraction conditions can be calculated numerically by combining Eq. (3.3) with the rigorous height difference function,

$$C_d(k_\perp, \mathbf{r}) = \frac{\sum_{n=-\infty}^{+\infty} e^{-\frac{1}{2} H_d(\mathbf{r}) (\phi - 2\pi n)^2}}{\sum_{n=-\infty}^{+\infty} e^{-\frac{1}{2} H_d(\mathbf{r}) (2\pi n)^2}}. \quad (2.22)$$

Appendix IIIA One-dimensional Restricted Markovian Chain Surface

The basic idea of this model surface was proposed by Lent and Cohen. [3.3]

1. *The surface model*

Consider a one-dimensional array of steps, as shown in Fig. ApIIIA.1, where steps occur only in one particular direction (x-direction) on a two-dimensional surface. Assume that the surface layers have a AAA stacking structure, i.e., atoms in different layers sit directly over each other. The lateral lattice constant is a and the single layer step height (layer spacing) is given by c. As shown in Fig. ApIIIA.1, both single-layer and multilayer steps are allowed to exist in this surface model. In addition, we assume the steps are restricted in N levels, i.e., the surface height can fluctuate within N levels. The surface atom which is located at the lth level, $z(x_n) = h(x_n)c = lc$, (l = 1, 2, ..., N) with the lateral site, $x_n = na$, ($n = 0, \pm 1, \pm 2, \ldots$) can be represented by the coordinates, $(na, lc) = (n, l)$.

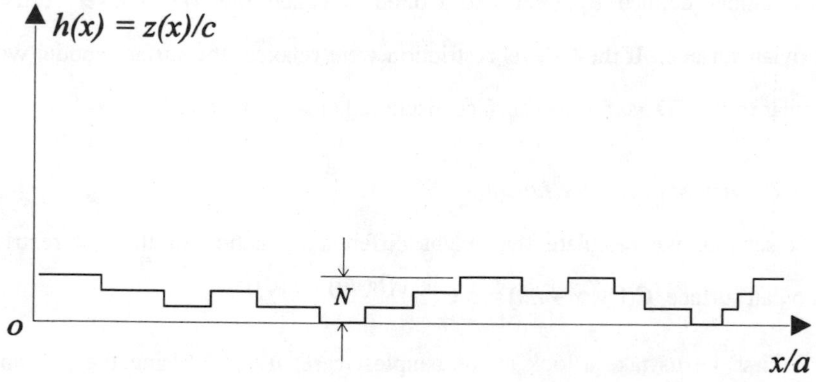

Fig. ApIIIA.1 Schematic of a restricted N-level one-dimensional stepped surface.

Let B_{lk} be the probability of meeting a surface position with the height displaced from the lth level to the kth level in going from any lattice site to its nearest neighbors, where $l, k = 1, 2, ..., N$. B_{lk} actually is the probability to meet a step at the lth level with the step height, $\Delta z = (k - l)c$, where $\Delta z > 0$ means an upward step and $\Delta z < 0$ refers to a downward step. The step probability B_{lk} is an element of the real $N \times N$ matrix,

$$B = (B_{lk}) = \begin{pmatrix} B_{11} & B_{12} & \dots & B_{1N} \\ B_{21} & B_{22} & \dots & B_{2N} \\ \dots & \dots & \dots & \dots \\ B_{N1} & B_{N2} & \dots & B_{NN} \end{pmatrix}. \qquad \text{(ApIIIA.1)}$$

The probability B_{lk} has the following normalized relationship,

$$\sum_{k=1}^{N} B_{lk} = 1, \quad (l = 1, 2, ..., N).$$

We also assume that each displacement or step occurs independent of the other under the condition that the displacement is within the N levels. The above surface model defined by Lent and Cohen is called the 1D N-level restricted Markovian surface. If the N-level restriction were relaxed, the surface model would be similar to the 1D Markovian surface discussed in Appendix VIA.

2. *The height difference function*

In this section, we calculate the height difference function in the 1D restricted Markovian surface, $C(k_\perp, x = na) = \langle e^{ik_\perp c[h(na) - h(0)]} \rangle$.

First, let us take a look at the simplest case, $n = 1$. Using the probability matrix defined by Eq. (ApIIIA.1), we have

$$C(k_\perp, a) = \langle e^{ik_\perp c[h(na) - h(0)]} \rangle = \sum_{k=1}^{N} \sum_{l=1}^{N} \theta_l B_{lk} e^{ik_\perp c(k-l)}, \qquad (\text{ApIIIA.2})$$

where $l = h(0)$ and $k = h(a)$. θ_l is the coverage of the surface atoms in the lth level, ($l = 1, 2, ..., N$). Since θ_l is equivalent to the probability of finding a surface atom in the lth level, it must have a normalized relationship, $\sum_{l=1}^{N} \theta_l = 1$.

Next, let us consider the case of $n \geq 2$. In the expression of $\langle e^{ik_\perp c[h(na) - h(0)]} \rangle$, the height difference, $[h(na) - h(0)]$, can be represented as the sum of the height differences between successive sites from $x = 0$ to $x = n$a,

$$h(na) - h(0) = [h(a) - h(0)] + [h(2a) - h(a)] +$$

$$+ [h(3a) - h(2a)] + ... + [h(na) - h((n-1)a)].$$

According to the definition of the independent step probability given by Eq. (ApIIIA.1), we can calculate the height difference function as

$$C(k_\perp, na) = \langle e^{ik_\perp c[h(na) - h(0)]} \rangle$$

$$= \langle e^{ik_\perp c\,[h(a) - h(0)]} \times e^{ik_\perp c\,[h(2a) - h(a)]} \times e^{ik_\perp c\,[h(3a) - h(2a)]} \times$$

$$\times ... \times e^{ik_\perp c\,[h((n-1)a) - h((n-2)a)]} \times e^{ik_\perp c\,[h(na) - h((n-1)a)]} \rangle$$

$$= \sum_{l_n=1}^{N} \sum_{l_{n-1}=1}^{N} \cdots \sum_{l_2=1}^{N} \sum_{l_1=1}^{N} \sum_{l=1}^{N} \theta_l \,[B_{ll_1} e^{ik_\perp c(l_1-l)}] \times [B_{l_1l_2} e^{ik_\perp c(l_2-l_1)}] \times$$

$$\times [B_{l_2l_3} e^{ik_\perp c(l_3-l_2)}] \times ... \times [B_{l_{n-2}l_{n-1}} e^{ik_\perp c(l_{n-1}-l_{n-2})}] \times [B_{l_{n-1}l_n} e^{ik_\perp c(l_n-l_{n-1})}],$$

where $l = \mathrm{h}(0)$, $l_1 = \mathrm{h}(a)$, $l_2 = \mathrm{h}(2a)$, ..., $l_{n-1} = \mathrm{h}((n-1)a)$, $l_n = \mathrm{h}(na)$. The expression of the height difference function shown above can be further simplified to

$$\mathrm{C}(\mathrm{k}_\perp, na) = \sum_{l_n=1}^{\mathrm{N}} \sum_{l_{n-1}=1}^{\mathrm{N}} \cdots \sum_{l_2=1}^{\mathrm{N}} \sum_{l_1=1}^{\mathrm{N}} \sum_{l=1}^{\mathrm{N}} \theta_l \, e^{i\mathrm{k}_\perp c(l_n - l)} [\mathrm{B}_{ll_1} \mathrm{B}_{l_1 l_2} \cdots \mathrm{B}_{l_{n-2} l_{n-1}} \mathrm{B}_{l_{n-1} l_n}]$$

$$= \sum_{k=1}^{\mathrm{N}} \sum_{l=1}^{\mathrm{N}} \theta_l \, e^{i\phi(k-l)} (\mathrm{B}^n)_{lk} , \qquad (\text{ApIIIA.3})$$

where $\phi = \mathrm{k}_\perp c$ and we replace l_n by k. B^n is the product of n number of the matrix B. If $n = 1$, Eq. (ApIIIA.3) recovers Eq. (ApIIIA.2).

Suppose we can solve the eigenvalue equation of the matrix B,

$$\det[\mathrm{B} - \lambda] = 0 ,$$

where λ is the diagonalized matrix,

$$\lambda = \begin{pmatrix} \lambda_1 & 0 & \ldots & 0 \\ 0 & \lambda_2 & \ldots & 0 \\ \ldots & \ldots & \ldots & \ldots \\ 0 & 0 & \ldots & \lambda_\mathrm{N} \end{pmatrix} ,$$

in which, λ_l is the lth eigenvalue of the matrix B, ($l = 1, 2, ..., \mathrm{N}$). In principle, one can find a diagonalizing matrix Q of the matrix B,

$$\mathrm{Q}^{-1} \mathrm{B} \, \mathrm{Q} = \lambda \;\text{ or }\; \mathrm{B} = \mathrm{Q} \, \lambda(\phi) \, \mathrm{Q}^{-1} , \qquad (\text{ApIIIA.4})$$

where Q^{-1} is the reverse matrix of Q, $\mathrm{QQ}^{-1} = \mathrm{Q}^{-1}\mathrm{Q} = \mathrm{I}$. The matrix elements Q_{lk} can be obtained from the linear equations,

$$\sum_{k=1}^{N} (B_{lk} - \lambda_j \delta_{lk}) Q_{kj} = 0, \qquad (l, j = 1, 2, ..., N).$$

Combining Eq. (ApIIIA.3) with Eq. (ApIIIA.4), we have

$$C(k_\perp, na) = \sum_{k=1}^{N} \sum_{l=1}^{N} \theta_l \, e^{i\phi(k-l)} \, (Q \lambda^n Q^{-1})_{lk}$$

$$= \sum_{k=1}^{N} \sum_{l=1}^{N} \sum_{j=1}^{N} \theta_l \, e^{i\phi(k-l)} \, Q_{lj} \, \lambda_j^n \, Q_{jk}^{-1}.$$

$$= \sum_{j=1}^{N} \left(\sum_{k=1}^{N} \sum_{l=1}^{N} \theta_l \, Q_{lj} Q_{jk}^{-1} \, e^{i\phi(k-l)} \right) \lambda_j^n .$$

If we define the coefficient,

$$A_j(\phi) = \sum_{k=1}^{N} \sum_{l=1}^{N} \theta_l \, Q_{lj} Q_{jk}^{-1} \, e^{i\phi(k-l)}, \qquad (\text{ApIIIA.5})$$

we then obtain a simple expression of the height difference function,

$$C(k_\perp, na) = \sum_{j=1}^{N} A_j(\phi) \, \lambda_j^n . \qquad (\text{ApIIIA.6})$$

3. *The eigenvalues λ_j in the reversible surface*

The surface which has a symmetry between $+x$ and $-x$ directions is called the reversible surface, according to the original definition by Lent and Cohen. In the present problem, we only concentrate on the reversible surface. The major properties of the eigenvalues λ_j can be derived as follows.

(i) All eigenvalues are real

The condition of the reversibility leads to the following identity,

$$\theta_l \, B_{lk} = \theta_k \, B_{kl} \,, \qquad \text{(ApIIIA.7)}$$

which is the probability of finding two nearest neighbor atoms respectively at the lth and kth levels. Consider the matrix Γ defined as $\Gamma_{lk} = \theta_l^{1/2} B_{lk} \, \theta_k^{-1/2}$, i.e., $\Gamma = \Theta^{1/2} B \Theta^{-1/2}$, where

$$\Theta = \begin{pmatrix} \theta_1 & 0 & \dots & 0 \\ 0 & \theta_2 & \dots & 0 \\ \dots & \dots & \dots & \dots \\ 0 & 0 & \dots & \theta_N \end{pmatrix}.$$

Γ must have the same eigenvalues as that of matrix B because according to Eq. (ApIIIA.4), we have

$$\lambda = Q^{-1} B \, Q = (Q^{-1} \Theta^{-1/2}) \, (\Theta^{1/2} B \Theta^{-1/2}) \, (\Theta^{1/2} Q) = (\Theta^{1/2} Q)^{-1} \, \Gamma \, (\Theta^{1/2} Q),$$

i.e., the matrix Γ can be diagonalized to λ by the similarity transformation matrix, $\Theta^{1/2}Q$.

We note that the matrix Γ is symmetric, $\Gamma_{lk} = \Gamma_{kl}$, because

$$\Gamma_{lk} = \theta_l^{1/2} B_{lk} \, \theta_k^{-1/2} = \theta_l^{-1/2} \, (\theta_l \, B_{lk}) \, \theta_k^{-1/2}$$

$$= \theta_l^{-1/2} (\theta_k \, B_{kl}) \, \theta_k^{-1/2} = \theta_k^{1/2} B_{kl} \, \theta_l^{-1/2} = \Gamma_{kl} \,,$$

where we have used Eq. (ApIIIA.7). Since Γ is a symmetric matrix, all the eigenvalues λ_j must be real.

(ii) All eigenvalues are in the range, $-1 \le \lambda_j \le 1$, ($l = 1, 2, ..., N$).

Since $| < e^{ik_\perp c[h(na) - h(0)]} > | \le 1$, Eq. (ApIIIA.6) must satisfy

$$|C(k_\perp, na)| = |\sum_{j=1}^{N} A_j(\phi)\, \lambda_j^n| \le 1.$$

At $n \to \infty$, the above inequality can be held only under the condition, $|\lambda_j| \le 1$, ($j = 1, 2, ..., N$). We therefore conclude that all eigenvalues must be in the range, $-1 \le \lambda_j \le 1$.

(iii) There exists one but only one eigenvalue which is equal to unity.

As $n \to \infty$, the height difference function should become

$$C(k_\perp, na) = < e^{ik_\perp c[h(na) - h(0)]} >$$

$$\approx < e^{i\phi h(na)} > < e^{-i\phi h(0)} >$$

$$= \left(\sum_{k=1}^{N} \theta_k\, e^{i\phi k} \right) \left(\sum_{l=1}^{N} \theta_l\, e^{-i\phi l} \right)$$

$$= \sum_{k=1}^{N} \sum_{l=1}^{N} \theta_l\, e^{i\phi(k-l)}\, \theta_k \,, \; (n \to \infty). \qquad \text{(ApIIIA.8)}$$

Comparing the above equation with Eq. (ApIIIA.3), we immediately obtain, $(B^n)_{lk} = \theta_k$, as $n \to \infty$. Therefore, $\sum_{k=1}^{N} (B^n)_{kk} = \sum_{k=1}^{N} \theta_k = 1$, $(n \to \infty)$. Recall that $\sum_{k=1}^{N} (B^n)_{kk}$

is the trace of the matrix B^n, which should be related to the eigenvalues of B through the following identity,

$$\sum_{k=1}^{N} (B^n)_{kk} = \sum_{k=1}^{N} \lambda_k^n.$$

We thus obtain an asymptotic identity for the eigenvalues,

$$\sum_{k=1}^{N} \lambda_k^n = 1, \quad (n \to \infty).$$

This asymptotic identity indicates that there is only one eigenvalue which equals unity. From the asymptotic identity shown above, we also conclude that -1 is not an eigenvalue of the matrix B.

4. *The height difference function in the reversible surface*

From the discussion of the last section, we know that the matrix B has only one eigenvalue equal to unity, $\lambda_N = 1$, (defined as the Nth eigenvalue). The remaining (N – 1) eigenvalues are real with the range given by $-1 < \lambda_j < 1$, ($j = 1, 2, ..., N - 1$). It can be shown that the existence of the negative eigenvalues, $-1 < \lambda_j < 0$, is the indication that the surface has certain reconstructed local portions in which the periodicity has changed from a to 2a. In this monograph, we do not discuss these kinds of reconstructed surface structures. For the non-reconstructed and reversible 1D restricted Markovian surface, we have

$$\lambda_j : \begin{cases} \lambda_N = 1 \\ 0 \le \lambda_j < 1, \ (j = 1, 2, ..., N - 1) . \end{cases}$$

Since there is no negative eigenvalue, we can rewrite the height difference function given by Eq. (ApIIIA.6) as

$$C(k_\perp, na) = \sum_{j=1}^{N} A_j(\phi)\, e^{-\omega_j na}$$

$$= A_N(\phi) + \sum_{j=1}^{N-1} A_j(\phi)\, e^{-\omega_j na}, \quad (n \geq 0), \qquad \text{(ApIIIA.6')}$$

where $\omega_j = -\frac{1}{a}\,\mathrm{Ln}(\lambda_j) > 0$ since $0 \leq \lambda_j < 1$, (j = 1, 2, ..., N − 1). The coefficients $A_j(\phi)$ can be considered as follows.

(i) The coefficient corresponding to the eigenvalue $\lambda_N = 1$

The coefficient $A_N(\phi)$ corresponding to the eigenvalue $\lambda_N = 1$ can be obtained from Eq. (ApIIIA.8), i.e.,

$$A_N(\phi) = \sum_{k=1}^{N}\sum_{l=1}^{N} \theta_l\, e^{i\phi(k-l)}\, \theta_k = \Big|\sum_{l=1}^{N} \theta_l\, e^{i\phi l}\Big|^2 \geq 0.$$

(ii) The coefficients for the eigenvalues $0 \leq \lambda_j < 1$

For the reversible surface which has a symmetry between + x and − x directions, the height difference function should have the following symmetric relation,

$$C(k_\perp, na) = C(k_\perp, -na).$$

Since $C(k_\perp, na) = \langle e^{\,ik_\perp c[h(na) - h(0)]} \rangle$, from $C(k_\perp, na) = C(k_\perp, -na)$, we have

$$C(k_\perp, na) = \langle e^{\,ik_\perp c[h(-na) - h(0)]} \rangle = \langle e^{\,ik_\perp c[h(0) - h(na)]} \rangle = [C(k_\perp, na)]^* .$$

Comparing above equation with Eq. (ApIIIA.6'), we obtain

$$A_j(\phi) = A_j(\phi)^*,$$

which indicates that all the coefficients, $A_j(\phi)$, (j = 1, 2, ..., N − 1) are real. In conclusion, the general height difference function can be expressed as

$$C(k_\perp, x) = A_N(\phi) + \sum_{j=1}^{N-1} A_j(\phi)\, e^{-\omega_j |x|}, \qquad (\text{ApIIIA.9})$$

where we set $x = \pm na$.

5. *The diffraction structure factor*

The calculation of the diffraction structure factor is straightforward. Using the continuous expression, Eq. (2.16), we obtain

$$S(k_\|, k_\perp) = 2\pi A_N(\phi)\, \delta(k_\|) + \sum_{j=1}^{N-1} 2A_j(\phi)\, \omega_j^{-1} L_1\left(\frac{k_\|}{\omega_j}\right), \qquad (\text{ApIIIA.10})$$

where $L_1(X)$ is the 1D Lorentzian function given by $L_1(X) = \dfrac{1}{1+X^2}$.

Equation (ApIIIA.10) indicates that the diffraction structure factor from a 1D restricted Markovian surface is the sum of a central δ-peak and a diffuse profile given by

$$S_{diff}(k_\|, k_\perp) = \sum_{j=1}^{N-1} 2A_j(\phi)\, \omega_j^{-1} L_1\left(\frac{k_\|}{\omega_j}\right).$$

The diffuse structure factor consists of (N−1) number of 1D Lorentzians, $\dfrac{1}{1+(k_\|/\omega_j)^2}$, with different widths, ω_j, (j = 1, 2, ..., N−1).

Appendix IIIB The Proof of An Asymptotic Identity [3.1]**:**

$$g(\Omega,\gamma)=\sum_{n=1}^{\infty}\frac{\Omega^n}{n!}n^{-\gamma}\to e^{\Omega}\,\Omega^{-\gamma},\quad \text{as } \Omega\to\infty. \qquad (\text{ApIIIB.1})$$

For γ = integer ≥ 0, the proof of Eq. (ApIIIB.1) is straightforward because as $\Omega \to \infty$,

$$\frac{g(\Omega,\gamma)}{e^{\Omega}\Omega^{-\gamma}}\to\frac{\frac{d}{d\Omega}g(\Omega,\gamma)}{\frac{d}{d\Omega}(e^{\Omega}\Omega^{-\gamma})}\to\frac{g(\Omega,\gamma-1)}{e^{\Omega}\Omega^{-(\gamma-1)}}\to\cdots\to\frac{g(\Omega,1)}{e^{\Omega}\Omega^{-1}}\to\frac{g(\Omega,0)}{e^{\Omega}}=\frac{[e^{\Omega}-1]}{e^{\Omega}}\to 1.$$

For the noninteger case, we consider the relation,

$$g(\Omega,\gamma)=\sum_{n=1}^{\infty}\frac{\Omega^n}{n!}n^{-\gamma}=\Omega+\Omega\sum_{n=1}^{\infty}\frac{\Omega^n}{n!}(n+1)^{-\gamma-1}. \qquad (\text{ApIIIB.2})$$

Using the identity,

$$\frac{1}{(n+1)^{\gamma+1}}=\frac{1}{n^{\gamma+1}(1+1/n)^{\gamma+1}}=\sum_{k=0}^{\infty}\frac{\Gamma(-\gamma)}{\Gamma(-\gamma-k)}n^{-(k+\gamma+1)}, \qquad (\text{ApIIIB.3})$$

we reexpress Eq. (ApIIIB.2) as

$$g(\Omega,\gamma)=\Omega+\Omega\sum_{k=0}^{\infty}\frac{\Gamma(-\gamma)}{\Gamma(-\gamma-k)}\left(\sum_{n=1}^{\infty}\frac{\Omega^n}{n!}n^{-(k+\gamma+1)}\right)$$

$$=\Omega+\Omega\sum_{k=0}^{\infty}\frac{\Gamma(-\gamma)}{\Gamma(-\gamma-k)}g(\Omega,k+\gamma+1), \qquad (\text{ApIIIB.4})$$

where the last step is obtained using the definition of $g(\Omega, \gamma)$. If Ω tends to infinity, Eq. (ApIIIB.4) has an asymptotic solution:

$$g(\Omega, \varepsilon) = \frac{F(\Omega)}{\Omega^{\varepsilon}}, \qquad \text{(ApIIIB.5)}$$

where $F(\Omega)$ is a function of Ω and satisfies

$$\frac{\Omega}{F(\Omega)\,\Omega^{-\gamma}} \rightarrow 0 \quad \text{as } \Omega \rightarrow \infty. \qquad \text{(ApIIIB.6)}$$

To prove that Eq. (ApIIIB.5) is an asymptotic solution, we insert Eq. (ApIIIB.5) into the right side of Eq. (ApIIIB.4) and let $\Omega \rightarrow \infty$. The verification is simple:

$$\text{Right hand side of Eq. (ApIIIB.4)} = \Omega + \frac{F(\Omega)\,\Omega}{(1+\Omega)^{\gamma+1}} \rightarrow \frac{F(\Omega)}{\Omega^{\gamma}} = g(\Omega, \gamma)$$

$$= \text{Left hand side of Eq. (ApIIIB.4)},$$

where we have used Eqs. (ApIIIB.3) and (ApIIIB.6). It is easy to determine the function $F(\Omega)$:

$$F(\Omega) = g(\Omega, 0) = \sum_{n=1}^{\infty} \frac{\Omega^n}{n!} = e^{\Omega} - 1 \rightarrow e^{\Omega}, \text{ as } \Omega \rightarrow \infty.$$

Therefore, we have proved that $g(\Omega, \gamma) = \dfrac{F(\Omega)}{\Omega^{\gamma}} = e^{\Omega}\,\Omega^{-\gamma}$.

Appendix IIIC The Proof of An Asymptotic Equation as $\Omega \to \infty$ [3.1]:

$$2\pi\xi^2 e^{-\Omega} \sum_{m=1}^{\infty} \frac{\Omega^m}{m!} \int_0^{\infty} x dx [1-f(x)]^m J_0(k_{||}\xi x) \to 2\pi(\xi\Omega^{-\frac{1}{2\alpha}})^2 F_\alpha(k_{||}\xi\Omega^{-\frac{1}{2\alpha}}), \qquad (\text{ApIIIC.1})$$

where the left hand side of Eq. (ApIIIC.1) is equal to $S_{diff}(\mathbf{k}_{||}, k_\perp)$ given by Eq. (3.9).

We can make a transformation in $S_{diff}(\mathbf{k}_{||}, k_\perp)$, $x \leftrightarrow z$, defined by

$$\begin{cases} e^{-z^{2\alpha}} = 1 - f(x) \\ x = Q(z) = f^{-1}(1 - e^{-z^{2\alpha}}), \end{cases} \qquad (\text{ApIIIC.2})$$

where $f^{-1}(y)$ is the inverse function of $y = f(x)$. The left hand side of Eq. (ApIIIC.1) then becomes

$S_{diff}(\mathbf{k}_{||}, k_\perp) =$

$$= \xi^2 \sum_{m=1}^{\infty} e^{-\Omega} \frac{1}{m!} \Omega^m m^{-\frac{1}{2\alpha}} \int_0^{\infty} dz e^{-z^{2\alpha}} Q(zm^{-\frac{1}{2\alpha}}) Q'(zm^{-\frac{1}{2\alpha}}) J_0[k_{||}\xi Q(zm^{-\frac{1}{2\alpha}})]. \qquad (\text{ApIIIC.3})$$

The function $Q(z)$ is expected to be an analytical function in the region, $0 \le z < +\infty$. We can therefore expand the product,

$$Q(\rho)\, Q'(\rho)\, J_0[\gamma Q(\rho)] = \sum_{n=0}^{\infty} C_n(\gamma)\, \rho^n, \qquad (\text{ApIIIC.4})$$

as a Taylor series in terms of ρ. Inserting Eq. (ApIIIC.4) into Eq. (ApIIIC.3) gives

$$S_{diff}(\mathbf{k}_{||}, k_\perp) = 2\pi\xi^2 \int_0^{\infty} dz e^{-z^{2\alpha}} \sum_{n=0}^{\infty} C_n(k_{||}\xi)\, z^n \left(e^{-\Omega} \sum_{m=1}^{\infty} \frac{\Omega^m}{m!} n^{-\frac{1+n}{\alpha}} \right). \qquad (\text{ApIIIC.5})$$

As shown in Appendix (IIIB), under the condition, $\Omega >> 1$, we have

$$e^{-\Omega} \sum_{m=1}^{\infty} \frac{\Omega^m}{m!} n^{-\frac{1+n}{\alpha}} \approx \Omega^{-\frac{1+n}{\alpha}}, \qquad (\Omega >> 1).$$

Using this asymptotic identity, we can simplify Eq. (ApIIIC.5):

$S_{diff}(\mathbf{k}_{\|}, k_{\perp}) =$

$$= 2\pi\xi^2 \int_0^{\infty} x dx[1 - f(x)]^{\Omega} J_0(k_{\|}\xi x) = 2\pi\xi^2 \int_0^{\infty} r dr[1 - f(\frac{r}{\xi})]^{\Omega} J_0(k_{\|} r). \quad \text{(ApIIIC.6)}$$

The scaling function $f(x)$ indicates that

$$1 - g(\frac{r}{\xi}) \sim \begin{cases} 1 - \left(\frac{r}{\xi}\right)^{2\alpha} & \text{for } r << \xi, \\ 0 & \text{for } r >> \xi. \end{cases}$$

Given the condition $\Omega >> 1$, we have

$$[1 - f(\frac{r}{\xi})]^{\Omega} = \exp\{\Omega \, \mathrm{Ln}[1 - g(\frac{r}{\xi})]\} \approx \begin{cases} e^{-\Omega(r/\xi)^{2\alpha}} & \text{for } r << \xi, \\ 0 & \text{for } r = \text{any other value}. \end{cases}$$

Therefore, only in the region, $0 \leq r << \xi$, can the function, $[1 - f(\frac{r}{\xi})]^{\Omega}$, have a significant contribution to the integral of Eq. (ApIIIC.6). We can thus replace in Eq. (ApIIIC.6) the function $[1 - f(\frac{r}{\xi})]^{\Omega}$ by $e^{-\Omega(r/\xi)^{2\alpha}}$, which yields Eq. (ApIIIC.1).

REFERENCES

3.1 H.-N. Yang, T.-M. Lu, and G.-C. Wang, *Phys. Rev. Lett.* **68**, 2612 (1992); *Phys. Rev.* **B47**, 3911 (1993).

3.2 S. K. Sinha, E. B. Sirota, S. Garoff, and H. B. Stanley, *Phys. Rev.* **B38**, 2297 (1987).

3.3 C. S. Lent and P. I. Cohen, *Surf. Sci.* **139**, 121 (1984).

3.4 M. Horn and M. Henzler, *J. of Crystal Growth* **81**, 428 (1987).

3.5 J. M. Bennett and L. Mattsson, *Introduction to Surface Roughness and Scattering* (Optical Society of America, Washington, D.C., 1989).

3.6 G. J. Pietsch, M. Henzler, and P. O. Hahn, *Appl. Surf. Science* **39**, 457 (1989).

Chapter IV GROWTH DYNAMICS: SELF-AFFINE SCALING

Growth morphologies and patterns are subjects of immense interest from both practical and fundamental point of views. Growth processes are inherently non-equilibrium processes. A standard statistical mechanics approach to a non-equilibrium process has not been developed so far and cannot be used to describe the growth dynamics and therefore the complex morphologies and patterns of the growth fronts. The dynamic scaling approach proposed in recent years is perhaps the most interesting and important theory developed in the area of non-equilibrium growth research.[4.1 - 4.3] In this Chapter, we shall outline the basic ingredience underlying the scaling hypothesis. The dynamic scaling properties of the height-height correlation function and the height difference function are presented.

§IV.1 The Dynamic Scaling Hypothesis

Surface roughening is a common phenomenon in far-from-equilibrium thin film growth. The interface morphology evolves in time during growth and it involves a complicated time-dependent dynamic roughening process. Theoretically, it has been generally recognized that the surface morphology and dynamics of a growing interface exhibit simple dynamic scaling behavior despite the complication of the growth processes. The dynamic scaling approach can significantly reduce the great number of degrees of freedom for the description of the interface morphology. In this section, we present an outline of this approach.

§IV.1.1 *Basic physics*

Thin film growth basically is a process of the addition or deposition of materials onto a surface. During the growth process, two competing processes, random fluctuation and local smoothing effects, play key roles in the evolution of the surface morphology. Random fluctuation, which can arise either from deposition rate or from thermal evaporation process, is the major cause of the height fluctuation and then the surface roughening during the growth. In contrast, the smoothing effects, such as thermal diffusion or side growth, are the relaxation mechanisms that try to eliminate the surface height fluctuation. Local smoothing effects (such as diffusion) always occur for small distances while the random fluctuation can take place in both small and large distances. Therefore, these two competing processes can only reach a balance on a relatively short-range scale but not on the long-range scale. This competition leads to the generic surface characteristics during dynamic growth.

The competition between fluctuations and smoothing eventually reaches a balance on a relatively short-range scale (shorter than the correlation length $\xi(t)$), so that the local surface morphology is statistically stationary (time-invariant). In spite of the short-range stationary morphology, the global surface morphology still evolves with time during growth. Shown in Fig. 4.1 is the STM image of a sputter-deposited Cr film on Si. [4.4] Figures 4.1(a), 4.1(b) and 4.1(c) illustrate the evolutions of a mountain-valley structure of the film at different deposition times. Initially, the mountain size (or grain size) is small, as shown in Fig. 4.1(a). The average size of the mountain is proportional to $\xi(t)$ and its average height is proportional to $w(t)$. As the deposition time increases, shown in Figs. 4.1(b) and 4.1(c), the grain size becomes larger and larger, i.e., the random fluctuations occur on a larger and larger scale in both the vertical (w(t)) and the lateral (ξ(t)) directions. Since the competition between fluctuations and smoothing does not reach a balance on the long-range scale, the global surface morphology thus proceeds to a steady state growth with the evolution of vertical roughening and lateral coarsening.

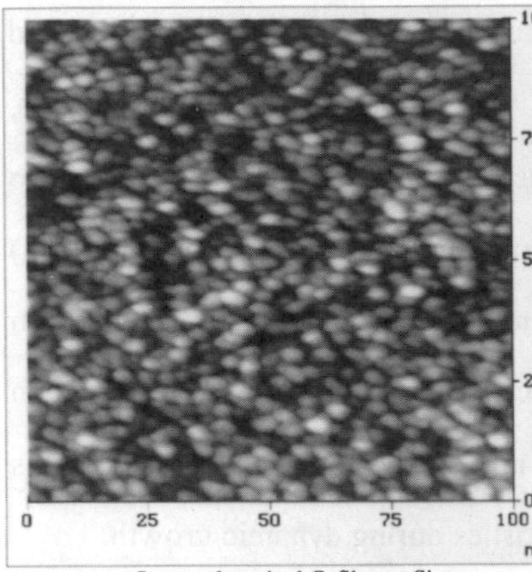

Sputter-deposited Cr film on Si,
p(Ar) = 20mTorr, time = 0.5 min.

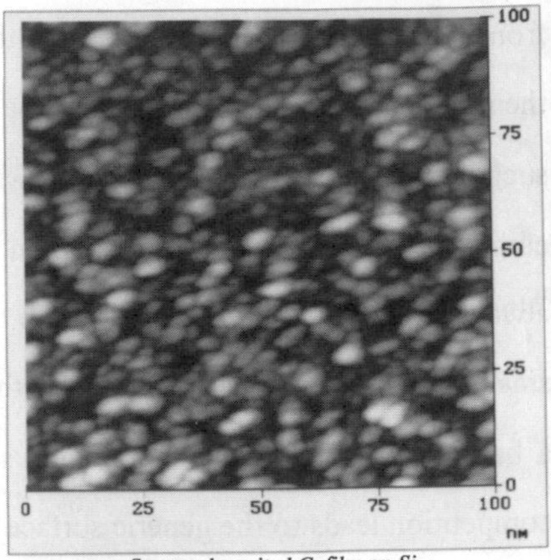

Sputter-deposited Cr film on Si,
p(Ar) = 20mTorr, time = 1.5 min.

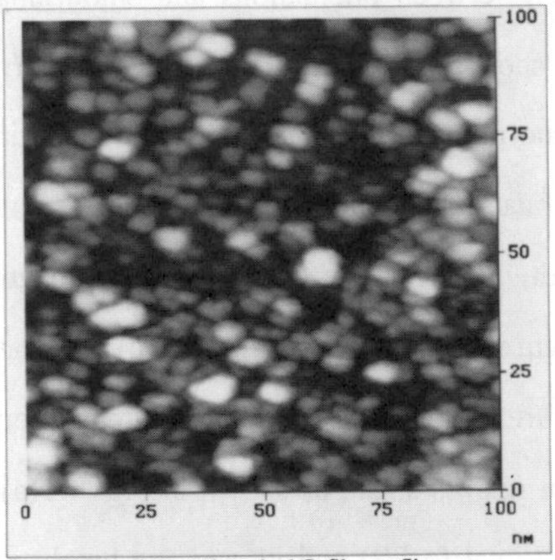

Sputter-deposited Cr film on Si,
p(Ar) = 20mTorr, time = 3.0 min.

Fig. 4.1 The evolution of the STM image for the thin film growth of sputter-deposited Cr on Si.[4.4]

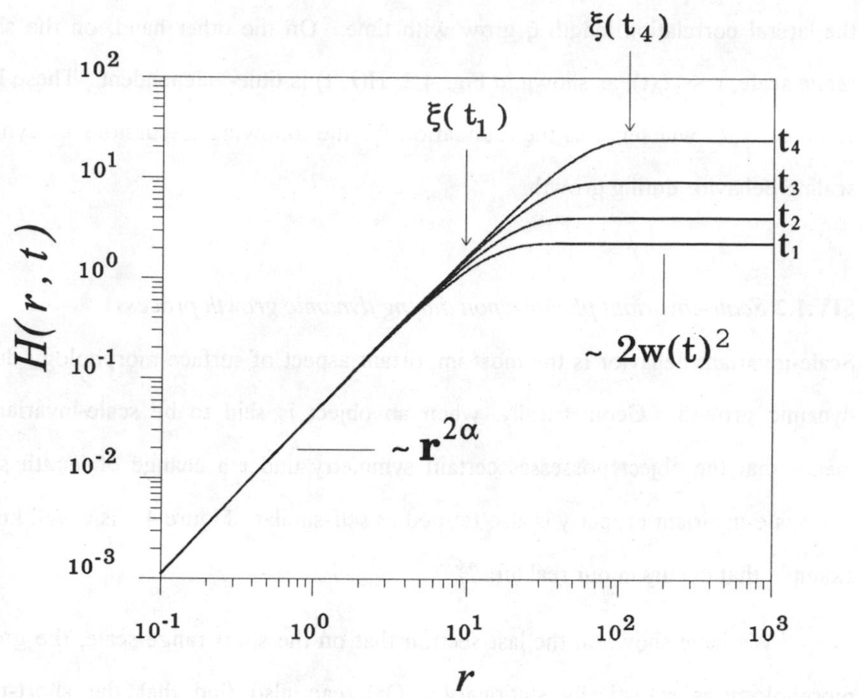

Fig. 4.2 The time-dependent height-height correlation function of an interface morphology during dynamic growth.[4,5] The lateral and vertical scales are in units of *a* and *c*, respectively.

In the above discussion, we have shown the possibility that both a short-range time-invariant state and a long-range time-dependent state can coexist during dynamic growth. To summarize the aspects discussed above, we illustrate in Fig. 4.2 a time-dependent height-height correlation function H(**r**, t) that describes the statistical evolution of the interface morphology during growth.[4,5] At the long-range scale, r >> ξ(t), the evolution of H(**r**, t) suggests that both the interface width w and the lateral correlation length ξ grow with time. On the other hand, on the short-range scale, $r << \xi(t)$, as shown in Fig. 4.2, $H(\mathbf{r}, t)$ is time-independent. These basic characteristics will serve as the foundation for the following discussions of dynamic scaling behavior during growth.

§IV.1.2 *Scale-invariant phenomenon during dynamic growth process*

Scale-invariant behavior is the most important aspect of surface morphology during dynamic growth. Geometrically, when an object is said to be scale-invariant, it means that the object possesses certain symmetry under a change of length scale. The scale-invariant property is also termed as self-similar. Figure 4.3 is a well known example that occurs in our real life. [4.6]

We have shown in the last section that on the short-range scale, the growth morphology is statistically stationary. One can also find that the short-range morphology is scale-invariant at different length scales. Shown in Fig. 4.4 is the images which are repeated magnifications of a local section from a surface having a cauliflower structure.[4.7] In this example, the scale-invariant phenomenon appears in the sense that if we expand a local section from a larger local section (both less than $\xi(t)$), the basic features of the surface are not changed. More rigorously, such a scale-invariant aspect is related to the self-affine characteristic, which will be discussed in §IV.1.3.

Also, the growth is scale-invariant in time. As shown in Fig. 4.1, the long-range topographies of Cr/Si films at different growth times have different characteristic grain size. Assume we can properly rescale the pictures of Figs. 4.1(a), 4.1(b) and 4.1(c) so that these three topographies have the same characteristic size of

4.1(b) and 4.1(c) so that these three topographies have the same characteristic size of grains. Practically, we can enlarge Fig. 4.1(a) by a certain amount and enlarge Fig. 4.1(b) by a different amount so that they can match the characteristic size of grains in Fig. 4.1(c). After these scaling processes, we will find that the rescaled morphologies for the different growth times are quite similar. From the statistical point of view, we can say that the rescaled structures are identical. The global surface morphology during growth thus evolves with time in a self-similar manner. This phenomenon has been observed in many growth experiments.

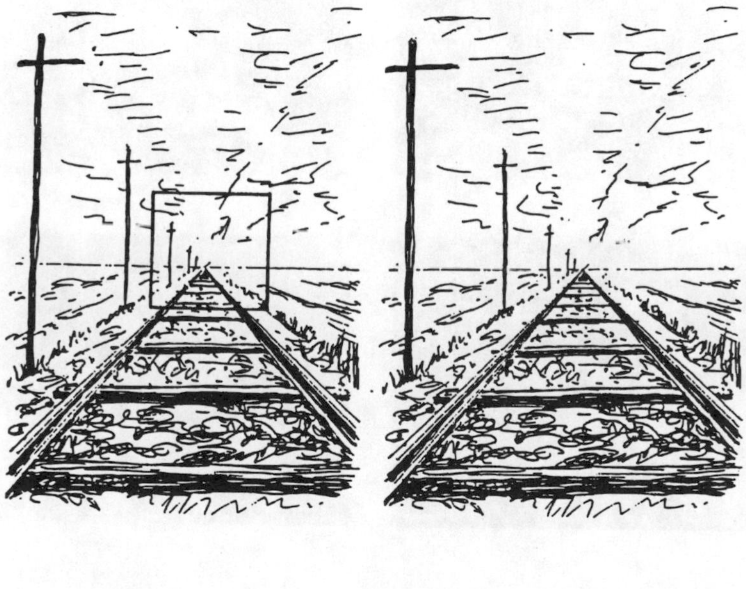

Fig. 4.3 A real-life example of scale-invariant behaviors. The picture in the right hand side is a magnification of the enclosed section in the left figure. [4.6]

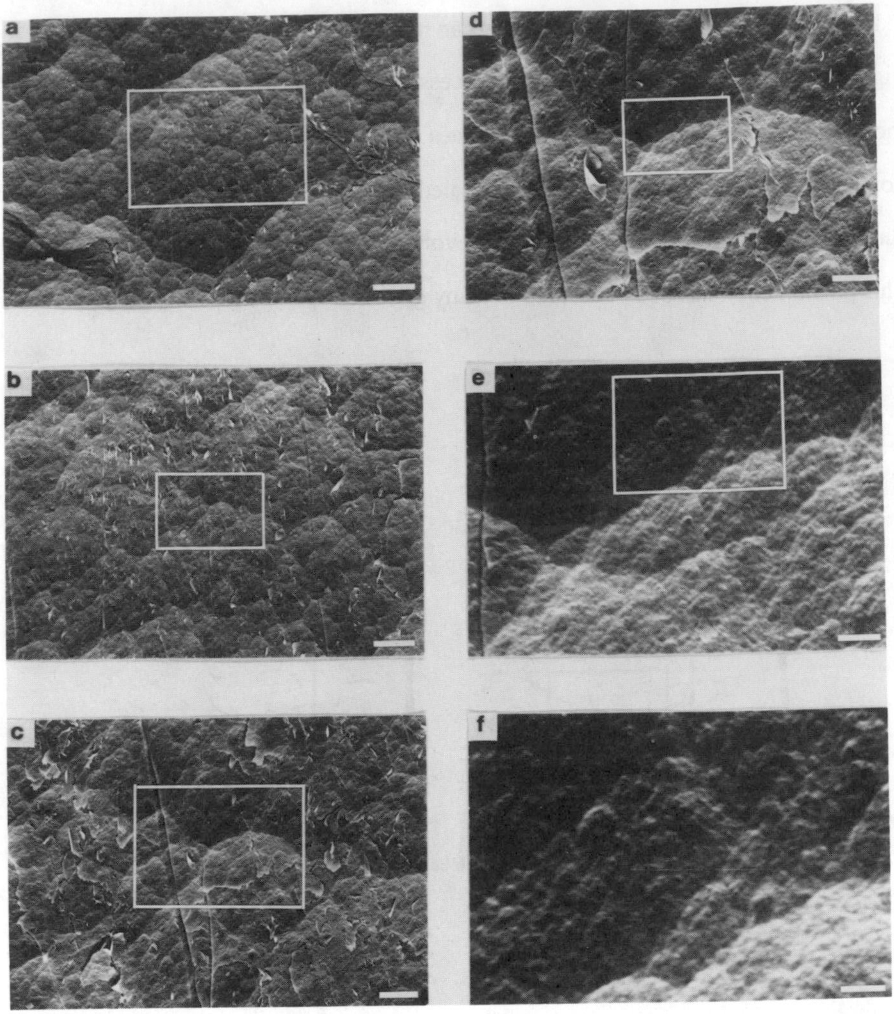

Fig. 4.4 Repeated magnification of a cauliflower structure (courtesy of R. Messier [4.7]).

§IV.1.3 *Dynamic scaling hypothesis in growing surfaces: a quantitative description*

The dynamic scaling hypothesis implies that the growing surface is scale-invariant both in time and space, where the invariant behavior refers to the statistical quantities from the surface. These statistical quantities are space scale-invariant at short distances ($< \xi(t)$) and time scale-invariant at large distances ($> \xi(t)$). We expect that any statistical quantity of the growing surface would have such scale-invariant characteristics.

Consider the interface width w of a growing surface,

$$w(L, t) = \sqrt{< (z - <z>_L)^2 >_L}\,, \qquad (4.1)$$

where L is the lateral linear size of the sections in the growing surface and $< >_L$ means the average over these sections. The interface width defined in Eq. (4.1) is slightly more general than the definition given in Chapter II. w(L, t) in Eq. (4.1) is measured over a finite-size regime with a lateral size L, which can vary from short-range to long-range distances. In contrast, w defined in the previous chapter refers to the long-range regime only, i.e., the global interface width. For the present growth problem, the global interface width can be defined as

$$w(t) = \lim_{L \to \infty} w(L, t). \qquad (4.2)$$

The dynamic scaling hypothesis proposes the following scale-invariant relation in a growing surface[4.1]: if one rescales the surface by a factor of λ in the lateral direction, in order to observe the similarity between the original and the rescaled surface, one has to not only rescale the surface by a factor of λ^{α} in the vertical direction, but also rescale the time by a factor of λ^{γ} corresponding to the time, $t' = \lambda^\gamma t$. Here, the value of the scaling factor λ is assumed to be $0 < \lambda < 1$. The scale-invariant relation can be summarized as

$$\begin{cases} \mathbf{r} \to \lambda \mathbf{r} & \text{(the lateral scale)} \\ z \to \lambda^\alpha z & \text{(the vertical scale)} \\ t \to \lambda^\gamma t & \text{(the time scale)} \end{cases} , \qquad (4.3)$$

where α and γ are the growth exponents. The exponent α is a measure of the surface roughness. The physical meaning of these exponents will be discussed in §IV.3. We can show later that the proposed scaling relation, Eq. (4.3), includes both the space scale-invariant and the time scale-invariant aspects in the growth front.

According to Eq. (4.3), the interface width defined in Eq. (4.1) should be scaled to give a generalized homogeneous function,

$$w(\lambda L, \lambda^\gamma t) = \lambda^\alpha w(L, t) . \qquad (4.4)$$

Let $\lambda = 1/L$. We obtain

$$w(L, t) = L^\alpha w(1, \frac{t}{L^\gamma}) = L^\alpha g\left(\frac{t}{L^\gamma}\right), \qquad (4.5)$$

where $g\left(\frac{t}{L^\gamma}\right) = w(1, \frac{t}{L^\gamma})$ is a function of (t/L^γ). The function $g(x)$, where $x = \frac{t}{L^\gamma}$, is called the scaling function. From the long-range and the short-range characteristics during the growth process, we can determine the asymptotic behavior of $g(x)$.

First, we consider L to be small, $L < \xi(t)$. We may recall the fact from §IV.1.1 that due to the competition between fluctuation and smoothing during growth, the short-range surface morphology can reach a statistically stationary state after a sufficiently long time ($t \to \infty$). For the interface width under consideration, we have

$$\frac{d}{dt} w(L, t) = 0, \quad \text{as } t \to \infty, \tag{4.6}$$

which reflects the fact that the interface width w(L, t) measured within small L is a time-invariant quantity. Inserting Eq. (4.5) into Eq. (4.6) yields $\frac{d}{dt} g\left(\frac{t}{L^{\gamma}}\right) = 0$, as $t \to \infty$, i.e., $\frac{d}{dx} g(x) = 0$, as $x \to \infty$. We thus obtain the asymptotic behavior,

$$g(x) = \text{constant, as } x \to \infty.$$

We now turn to the large L case, $L \to \infty$. For a sufficiently large system, $L \to \infty$, the measured interface width, which is the statistical average of global height fluctuation, should not be affected by finite-size effects or boundary conditions. In other words, the interface width w(L, t) is not L-dependent if L is sufficiently large. We then have

$$\frac{d}{dL} w(L, t) = 0, \quad \text{as } L \to \infty. \tag{4.7}$$

Inserting Eq. (4.5) into Eq. (4.7), one obtains a differential equation,

$$\alpha L^{\alpha-1} g\left(\frac{t}{L^{\gamma}}\right) - \gamma\, t\, L^{\alpha-\gamma-1} g'\left(\frac{t}{L^{\gamma}}\right) = 0, \text{ as } L \to \infty,$$

i.e.,

$$\frac{g'(x)}{g(x)} = \frac{\alpha}{\gamma}\frac{1}{x}, \text{ as } x \to 0, \tag{4.7'}$$

where $x = \frac{t}{L^{\gamma}}$. The solution of the differential Eq. (4.7') gives a power-law form for the scaling function on the long-range scale,

$$g(x) \propto x^{\beta}, \quad \text{as } x \to 0,$$

where the exponent β is defined as $\beta = \alpha/\gamma$, or $\gamma = \alpha/\beta$.

In conclusion, the asymptotic behavior of the scaling function g(x) from a growing surface can be derived to have a form,

$$g(x) \approx \begin{cases} x^{\beta} & \text{for } x \ll 1 , \\ \text{constant} & \text{for } x \gg 1 . \end{cases} \quad (4.8)$$

We can now show that the scaling form of the interface width w(L, t), as given by Eqs. (4.5) and (4.8), does include the basic aspects observed in the growth process described in §IV.1.1 and §IV.1.2.

(i) Local structure: the self-affine fractal

Consider the short-range regime, $L \ll t^{1/\gamma}$, i.e., $x = t/L^{\gamma} \gg 1$. The interface width from Eqs. (4.5) and (4.8) is then given by

$$w(L, t) = w(L) \sim L^{\alpha}, \qquad (L \ll t^{1/\gamma}) . \quad (4.9)$$

This indicates that the interface width in the short-range regime is a time-invariant quantity.

The short-range interface width shown in Eq. (4.9) has a scale-invariant form,

$$w(\lambda L) = \lambda^{\alpha} w(L), \quad (4.10)$$

which is consistent with the experimental observation shown in Fig. 4.4. Quantitatively, Eq. (4.10) shows an anisotropic scale-invariant behavior: different scaling relations exist in the vertical and lateral directions. If one rescales the surface by a factor of λ along the lateral direction, according to Eq. (4.10), one must rescale the surface in the vertical direction by a factor of λ^{α} in order to observe similarity of the morphologies between the original and the rescaled surfaces. This anisotropic scale-invariant behavior also quantitatively agrees with the experimental observations. For a growing surface, one can expect a clear distinction of the morphology characteristics between the growth direction (vertical direction) and the lateral direction. Therefore, it is not surprising that the surface morphology has different scaling properties along different directions. Such an "anisotropic" scale-invariant property is closely related to that of a self-affine fractal if $0 \leq \alpha < 1$.

One can give a formal definition of the self-affine fractal [4.8]: a *d*-dimensional object is called a self-affine fractal if its fractal dimension D is scale dependent given by

$$D = \begin{cases} d+1-\alpha & \text{for short length scale ,} \\ d & \text{for large length scale .} \end{cases} \qquad (4.11)$$

Detailed descriptions about fractal characteristics can be found in many books and publications. Here we only refer to some of the salient features. In Eq. (4.10), if L is large, we obtain $w(L)/L \sim L^{\alpha-1} \rightarrow 0$ because $0 \leq \alpha < 1$. Therefore, on length scales much larger than w(L), (but still in the short-range regime, i.e., $w << L << \xi$), the growing surface appears to be flat and is a two-dimensional object, $D = 2$. This is because the surface fluctuation amplitude w(L) is negligibly small compared with the length scale L. However, in Eq. (4.10), if $L \rightarrow 0$, we would obtain $w(L)/L \sim L^{\alpha-1} \rightarrow \infty$. This indicates that at very short distances, the surface morphology looks very steep and very rough. The surface roughness can thus be characterized by the exponent α, as is demonstrated in Chapter II. It has been shown that such kind of rough surface at $w(L) > L$ is a fractal with a fractal dimension, $D = 3 - \alpha$. (Interested readers are referred to Refs. 4.2, 4.3, 4.6, 4.9.)

In conclusion, the morphology of a growing surface described by Eq. (4.10) is consistent with a self-affine fractal defined in Eq. (4.11).

(ii) Global structure: roughening, coarsening and scale-invariant in time

Consider the long-range regime, $L \gg t^{1/\gamma}$, which leads to $x = t/L^{\gamma} \ll 1$. The interface width from Eq. (4.5) is given by

$$w(L, t) = w(t) \sim t^{\beta}, \qquad (L \gg t^{1/\gamma}). \tag{4.12}$$

In contrast to the local stationary interface width w(L), as shown in Eq. (4.10), the global roughness, w(t), grows with time in the form of a power law.

From the above analysis, we may notice that $t^{1/\gamma}$ is a crucial quantity which determines both the short-range scale ($L \ll t^{1/\gamma}$) and the long-range scale ($L \gg t^{1/\gamma}$), as shown in Eqs. (4.9) and (4.12). Recall that in the previous chapters, we use the lateral correlation length ξ to distinguish the short-range ($r \ll \xi$) and the long-range ($r \gg \xi$) regimes, as shown in Eq. (2.26). We can therefore conclude that $t^{1/\gamma}$ is directly proportional to the lateral correlation length in the growing interface,

$$\xi(t) \sim t^{1/\gamma} = t^{\beta/\alpha}. \tag{4.13}$$

Equations (4.12) and (4.13) indicate that the global surface morphology reaches a steady evolution of both the vertical roughening and the lateral coarsening in a form of power-laws. This is consistent with the qualitative prediction based on the growth competing mechanisms discussed in §IV.1.1. Consider the measurements of w(t) and $\xi(t)$ at two growth times, $t = t_1$ and $t = t_2$, where $t_1 < t_2$. At $t = t_2$, if we rescale the surface by the factor $\lambda = (t_1/t_2)^{\beta/\alpha}$ along the lateral direction and the factor $\lambda^{\alpha} = (t_1/t_2)^{\beta}$ along vertical direction, we obtain $\lambda^{\alpha}w(t_2) = w(t_1)$ and $\lambda\xi(t_2) = \xi(t_1)$, according to Eqs. (4.12) and (4.13), respectively. Note that this scale-invariant behavior is similar to self-affine relations for short-range morphology. The power-law forms, Eqs. (4.12) and (4.13), thus preserve the scale-invariant (in time) properties in the long-range regime, as discussed in §IV.1.2.

§IV.2 Correlation Functions during Dynamic Growth

The dynamic scaling characteristics shown above also manifest themselves in various statistical correlation functions.

§IV.2.1 *Equal time height-height correlation function during dynamic growth*

For time-dependent growth problems, we define an equal time height-height correlation function as

$$H(\mathbf{r}, t) = < [z(\mathbf{r}, t) - z(0, t)]^2 >,$$

where $z(\mathbf{r}, t)$, is the surface atomic height at the growth time t.

According to the dynamic scaling hypothesis presented in §IV.1.3, two correlation lengths along the vertical and lateral directions are assigned to describe the roughening and coarsening processes, respectively. The vertical correlation length, w(t), is a measure of the global interface width. The lateral correlation length, ξ(t), the distance over which surface fluctuations spread, characterizes the coarsening size at the growth time t. Both w(t) and ξ(t) evolve with time in the form of power laws, as given by Eqs. (4.12) and (4.13).

The height-height correlation function has a long-range asymptotic form, according to Eq. (2.24),

$$H(\mathbf{r}, t) \to 2[w(t)]^2 \sim t^{2\beta}, \qquad (r >> \xi(t)). \qquad (4.14)$$

On the other hand, the height-height correlation function has a short-range asymptotic form which reflects time-invariant and self-affine characteristics,

$$H(\mathbf{r}, t) \sim r^{2\alpha}, \quad (r << \xi(t)). \qquad (4.15)$$

This relationship is consistent with the expression of the short-range interface width shown in Eq. (4.9) where we set L = r.

The general dynamic scaling relationship given by Eq. (4.3) ensures the invariant statistical properties of a growing surface under the scale transformation. For the equal time height-height correlation function, it yields a relation,

$$H(\lambda\mathbf{r}, \lambda^{\alpha/\beta} t) = \lambda^{2\alpha} H(\mathbf{r}, t). \qquad (4.16)$$

In Eq. (4.16), let $\lambda = t^{-\beta/\alpha}$. One obtains a scaling equation,

$$H(\mathbf{r}, t) = t^{2\beta} H(\frac{\mathbf{r}}{t^{\beta/\alpha}}, 1). \qquad (4.17)$$

In order to be consistent with both the scaling equation, Eq. (4.17), and the asymptotic expressions, Eqs. (4.14) and (4.15), the equal time height-height correlation function for a growing interface must have a form of

$$H(\mathbf{r}, t) = 2[w(t)]^2 f\left(\frac{r}{\xi(t)}\right), \qquad (4.18)$$

with the scaling function, $f\left(\frac{r}{\xi(t)}\right) \propto H(\frac{r}{t^{\beta/\alpha}}, 1)$, given by

$$f(X) = \begin{cases} X^{2\alpha} & \text{for } X \ll 1, \\ 1 & \text{for } X \gg 1, \end{cases}$$

where $X \equiv r/\xi(t)$.

Equation (4.18) has the same form as that for a static rough surface given by Eq. (2.27). Such a form is the result of the self-affine scale-invariant properties for a growing surface. At the short-range regime, Eq. (4.18) becomes

$$H(\mathbf{r}, t) \approx 2\left(\frac{r}{\eta}\right)^{2\alpha}, \quad (r \ll \xi(t)), \qquad (4.15')$$

where $\eta = \xi(t)w(t)^{-\frac{1}{\alpha}}$ is a time-independent quantity because $\xi(t) \sim t^{\beta/\alpha}$ and $w(t) \sim t^{\beta}$. Such a short-range parameter can be interpreted as proportional to the average terrace size for a crystalline surface during growth.[4.5, 4.10] Thus, similar to the scaling expression for the interface width, as shown in Eq. (4.5), the scaling expression for the height-height correlation function of Eq. (4.18) addresses important issues in a dynamic growth process. Examples are the time-invariant and self-affine characteristics at the short-range regime (Eq. (4.15')) and the time-dependent power-law growth in the long-range regime (Eq. (4.14)). The scaling form of the height-height correlation function is also qualitatively consistent with the plot shown in Fig. 4.2.

§IV.2.2 *Height difference function during dynamic growth*

During dynamic growth, the time-dependent height difference function,

$$C(k_\perp, \mathbf{r}, t) = \langle e^{ik_\perp[z(\mathbf{r}, t) - z(0, t)]} \rangle, \qquad (4.19)$$

can be calculated by combining Eq. (4.18) either with Eq. (2.19) for a continuous surface or with Eq. (2.22) for a crystalline surface.

The dynamic scaling features are clearly shown in the evolution of the height difference function, $C(k_\perp, \mathbf{r}, t)$, as a function of time. For a given equal time height-height correlation function for different w(t) at different growth times, as shown in Fig. 4.5(a), one can plot the corresponding $C(k_\perp, \mathbf{r}, t)$ in Figs. 4.5(b) and 4.5(c), respectively, at $\phi = \pi$, the out-of-phase diffraction condition and at $\phi = 0.1\pi$, the near in-phase condition. The calculation employs the discrete form given by Eq. (2.22), corresponding to a crystalline interface growth.

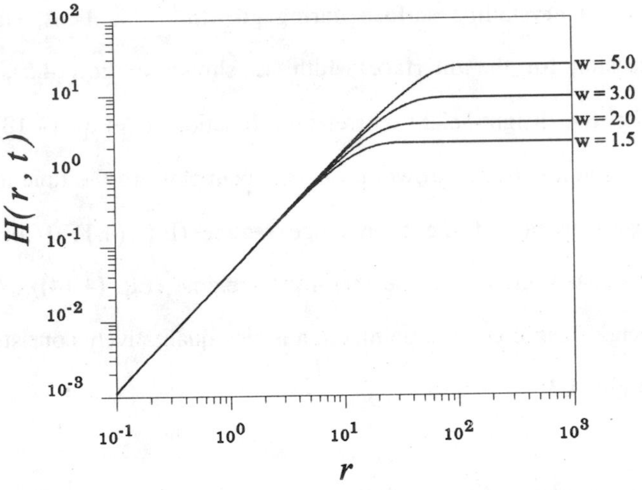

Fig. 4.5(a) Schematic of the equal time height-height correlation function, where the roughness exponent $\alpha = 0.8$ and the average terrace size, $\eta = 7.0$. The different values of w are the interface widths at different growth times. The relation between w and ξ is $\eta = \xi w^{-\frac{1}{\alpha}}$. The lateral and vertical scales are in units of *a* and *c*, respectively.

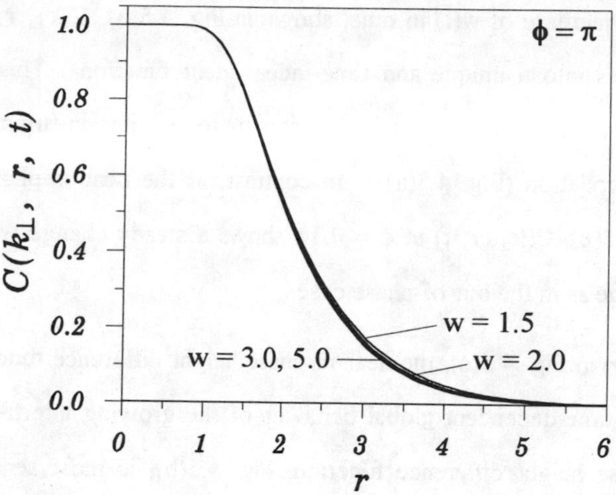

Fig. 4.5(b) Schematic of the height difference function corresponding to Fig. 4.5(a) at $\phi = \pi$, the out-of-phase diffraction condition.

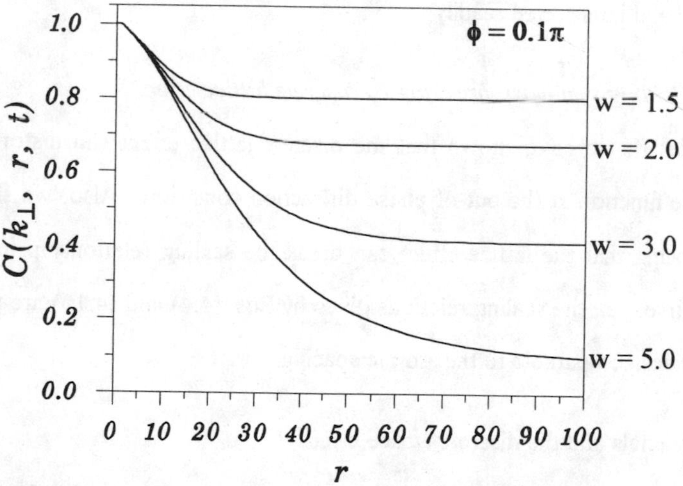

Fig. 4.5(c) Schematic of the height difference function corresponding to Fig. 4.5(a) at $\phi = 0.1\pi$, the near in-phase diffraction condition.

With the increase of w(t) in time, shown in Fig. 4.5(b), $C(k_\perp, \mathbf{r}, t)$ at $\phi = \pi$ quickly converges into a unique and time-independent function. This function is confined in a short-range regime which corresponds to the time-invariant part of the height-height correlation (Fig. 4.5(a)). In contrast, at the near in-phase condition shown in Fig. 4.5(c), $C(k_\perp, \mathbf{r}, t)$ at $\phi = 0.1\pi$ shows a steady change with time and does not converge as in the out-of-phase case.

As shown in Fig. 4.5(c), the near in-phase height difference function is most sensitive to the time-dependent global behavior of the growing interface while the near out-of-phase height difference function, Fig. 4.5(b), is most sensitive to the time-invariant local properties. Going from the in-phase to the out-of-phase conditions, the height difference function exhibits different scaling features. This very important result allows one to separate the time-invariant part from the time-dependent part in a diffraction experiment, so that the scaling properties can be measured and interpreted readily.

§IV.2.3 *Scaling symmetry-breaking by discrete lattice effect*

In Chapter II, we have shown that the discrete lattice effect can distort the height difference function at the out-of-phase diffraction condition. Also, we shall show in the following that the lattice effect can break the scaling relationship in the atomic scale regime, i.e., the scaling relations given by Eqs. (4.4) and (4.16) are not valid for length scales comparable to the atomic spacing.

(i) Fractals and the discrete lattice effect

In §IV.1.3, we have shown that on the short length scale, $L \ll r_c$, the growing surface morphology is a "$D = 3 - \alpha$" fractal, where r_c satisfies $w(r_c) \sim r_c$ or

$$w(r_c) \sim \sqrt{H(r_c, t)} \sim r_c .$$

According to Eq. (4.15'), we have $r_c \sim \eta^{-\alpha/(1-\alpha)}$. However, a growing surface in general is not a fractal on the atomic scale even though Eqs. (4.9) and (4.15') demonstrate a self-affine behavior. In the absence of surface overhangs, the average domain size, $\eta > 1$, (in units of the lateral atomic spacing *c*). Therefore, the fractal regime, $L << r_c \sim 1$, is seemingly cut off by the lattice constant because the scaling behavior, Eqs. (4.9) and (4.15'), does not exist within an atomic spacing. This discrete lattice effect has a significant impact on the scaling behavior in the diffraction measurements, which will be discussed later.

(ii) Scaling symmetry-breaking by discrete lattice effect

As shown above, the self-affine scaling relation does not apply to the atomic-scale regime, $L \sim 1$. This discrete lattice effect has a significant consequence on the height difference function. In the continuous surface model, according to Eq. (2.19), we have

$$C_c(k_\perp, \mathbf{r}, t) = e^{-\frac{1}{2}(k_\perp)^2 H_c(\mathbf{r}, t)} = e^{-[k_\perp w(t)]^2 f[r/\xi(t)]}, \qquad (4.20)$$

which is a self-affine scale-invariant function,

$$C_c(\lambda^{-\alpha}k_\perp, \lambda\mathbf{r}, \lambda^{\alpha/\beta}t) = C_c(k_\perp, \mathbf{r}, t) .$$

However, for a crystalline surface, if we insert Eq. (4.18) into Eq. (2.22), a direct examination of the height difference function shows that the self-affine scale-invariant relation usually does not exist, i.e.,

$$C_d(\lambda^{-\alpha}k_\perp, \lambda\mathbf{r}, \lambda^{\alpha/\beta}t) \neq C_d(k_\perp, \mathbf{r}, t), \qquad (4.21)$$

where

$$C_d(k_\perp, \mathbf{r}, t) \approx \frac{\sum\limits_{n=-\infty}^{+\infty} e^{-w(t)^2 f[\mathbf{r}/\xi(t)]\,(\phi - 2\pi n)^2}}{\sum\limits_{n=-\infty}^{+\infty} e^{-w(t)^2 f[\mathbf{r}/\xi(t)]\,(2\pi n)^2}}. \qquad (4.22)$$

The non-scaling behavior shown in Eq. (4.21) even persists in the lowest order expression,

$$C_0(k_\perp, \mathbf{r}, t) \approx e^{-\frac{1}{2}[\phi]^2 H_d(\mathbf{r})} = e^{-w(t)^2([\phi])^2 f[\mathbf{r}/\xi(t)]}, \qquad (4.23)$$

except the case of the "small $k_\perp$" diffraction condition. At the "small $k_\perp$" diffraction condition, $|\phi| << \pi$, since the difference between the continuous form and the discrete form can be ignored, we have $C_d(k_\perp, \mathbf{r}, t) \approx C_c(k_\perp, \mathbf{r}, t)$. As an example of the non-scaling behavior, we can directly examine $C_0(k_\perp, \mathbf{r}, t)$ for large ϕ under the scale transformation given by Eq. (4.3). From Eq. (4.23), we have

$$C_0(\lambda^{-\alpha}k_\perp, \lambda\mathbf{r}, \lambda^{\alpha/\beta}t) = e^{-w(t)^2(\lambda^{\alpha}[\lambda^{-\alpha}\phi])^2 f[\mathbf{r}/\xi(t)]},$$

where the calculation is based on Eqs. (4.12), (4.13) and $\phi = k_\perp c$ (c = vertical spacing). The scale-invariant relation,

$$C_0(\lambda^{-\alpha}k_\perp, \lambda\mathbf{r}, \lambda^{\alpha/\beta}t) = C_0(k_\perp, \mathbf{r}, t) = e^{-w(t)^2([\phi])^2 f[\mathbf{r}/\xi(t)]},$$

requires the following equality,

$$\lambda^{\alpha}[\lambda^{-\alpha}\phi] = \pm\,[\phi].$$

Such an equality can only hold in the case of $\phi \to 0$ or $|\phi| << \pi$, where $|\lambda^{-\alpha}\phi| \leq \pi$ so that $[\lambda^{-\alpha}\phi] = \lambda^{-\alpha}\phi$. For any other case, the equality is usually invalid. For example, if $\phi = 3\pi/5$, $\lambda = 0.25$ and $\alpha = 0.5$, we have $[\lambda^{-\alpha}\phi] = [2\times3\pi/5] = -4\pi/5$, i.e.,

$$\lambda^{\alpha}[\lambda^{-\alpha}\phi] = -2\pi/5 \neq \pm 3\pi/5 = \pm [\phi].$$

This indicates a non-scaling characteristic when the value of $\phi = k_{\perp}c$ is not sufficiently small.

In conclusion, the self-affine scale-invariant relation can only exist in the "small $k_{\perp}$" regime which corresponds to a large length scale in the real space. On the contrary, in the "large $k_{\perp}$" regime which corresponds to a small length scale in the real space, the self-affine relation is broken by the discrete atomic structure of the surface. It is the "discrete lattice effect" that leads to the failure of the self-affine scale-invariant description for the height difference function. However, as we show later, such a symmetry-breaking effect does not prevent us from extracting the roughness exponent α and other important dynamic scaling relations in a diffraction experiment.

§IV.3 Universality of Dynamic Growth

The dynamical scaling description of the surface morphology shown above has been shown to be a successful approach for various growth processes. Within the dynamical scaling approach summarized above, different growth processes would give different values of the growth exponents, α and β. The exponents depend on general features of the system such as the dimensionality, the growth mechanism, the symmetries of the steady state, the symmetries of the equation of motion, and the conservation laws. They do not depend on the detailed form of atomic interactions.

These exponents may be calculated from highly idealized theoretical deposition models which retain the essential features of the growth process. The values of α and β can thus be used to classify the corresponding growth models.

Depending on the volume conservation law during growth, the growth models that are of practical importance can be classified into two types: the non-conservative and the conservative growth processes. We now briefly discuss these two types of growth processes.

(i) Non-conservative growth process

In the non-conservative growth process, such as in the Eden model [4.11 - 4.14] or the ballistic deposition, [4.15, 4.16] side growth is allowed with the creation of voids and overhangs, but the relaxation mechanism (desorption or diffusion) is not strong enough to eliminate these defects completely.

For simplicity, let us briefly introduce the ballistic deposition model. An example of the film obtained by a ballistic deposition is shown in Fig. 4.6. In the surface regime, there are overhangs which may be due to the existence of the dangling bonds among atoms. Deep in the bulk there are vacancies or holes originating from overhangs during growth. The formation of the overhangs and voids are related to the ballistic growth mechanism. In Fig. 4.7, we show schematically in atomic scale how overhangs are formed during growth. In ballistic deposition, particles rain down onto the substrate following straight-line trajectories in the columns in which they were dropped until they first encounter a particle in the deposit. The particles can sometimes stick to the sides of the deposit and then cause side-ways growth (or side growth). The side growth thus leads to the formation of the overhangs and voids. The ballistically grown film has a larger volume (or lower density) than that of the completely compact solid for the deposited materials. The ballistic growth model is therefore a non-conservative process in the sense that the volume of the growing film is larger than that of the compact film of the same amount of deposited material.

The non-conservative growth can be described analytically by a non-linear Langevin equation (known as the KPZ equation which was first proposed by Kardar, Parisi, and Zhang [4.17]). The coupling of the non-linear term in the KPZ equation represents the side-growth effect during the deposition process. For a strong coupling, exact results of $\alpha = \frac{1}{2}$ and $\beta = \frac{1}{3}$ can be obtained from the KPZ equation for one-dimensional system, which have also been confirmed by numerical simulations. For two-dimensional system, only the scaling relation $\alpha + \gamma = 2$ is known, but simulations also suggested that $\alpha = \frac{1}{2}$ and $\beta = \frac{1}{3}$.

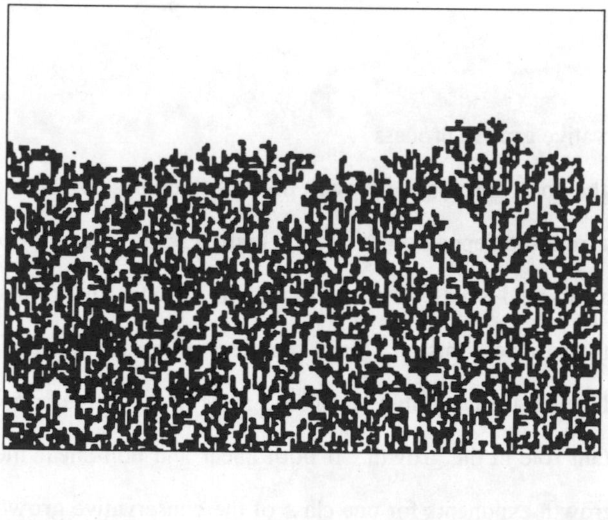

Fig. 4.6 Simulation of the interface morphology in ballistic deposition. [4.16]

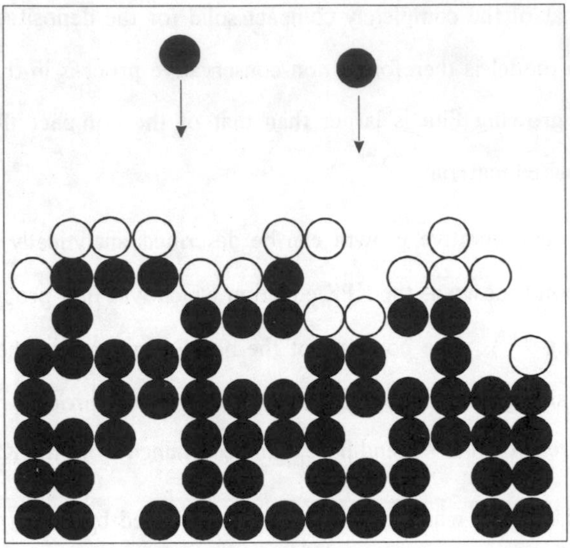

Fig. 4.7 Deposition geometry in ballistic growth. The open circles are the surface atomic sites. The holes in the interior are voids created during growth.

(ii) Conservative growth process

In contrast to the non-conservative growth process, the conservative growth creates a compact film from the deposition.[4.18 - 4.22] Figure 4.8 shows a compact grown film with no overhangs and voids. For conservative growth, the primary relaxation mechanism is the surface diffusion. Since the desorption of atoms and formation of overhangs and voids are negligibly small, the mass and volume conservation laws play an important role in the growth. If both linear and non-linear mechanisms are included, the growth exponents for one class of the conservative growth are $\alpha = 2/3$

and $\beta = 1/5$ for a $d = 2 + 1$ system [4.20] and $\alpha = 1$ and $\beta = \frac{2}{5}$ for a $d = 1 + 1$ system. [4.18] The values may vary depending on the couplings with other effects.

The roughness exponent α for conservative growth processes has larger values than that for non-conservative growth processes. As we have shown in Chapter II, smaller α gives a more jagged local surface morphology while larger α corresponds to a smoother local structure. In the conservative growth process, the significant diffusion effect creates smoother local structure in the surface. While in the non-conservative growth process, since the side growth creates overhangs and voids, the local surface structure tends to be very jagged.

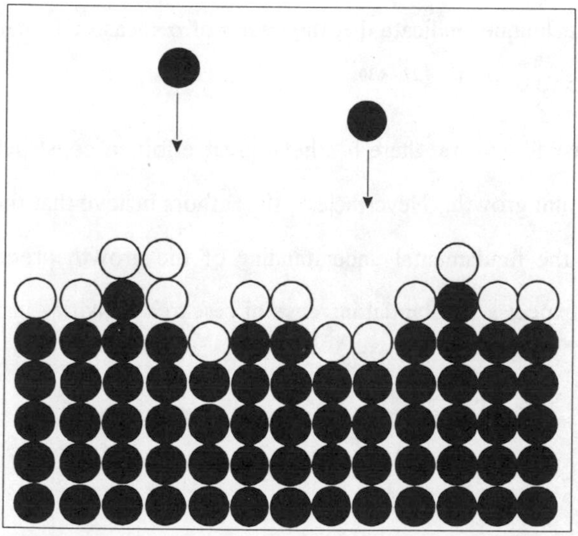

Fig. 4.8 Deposition geometry in conservative growth. The open circles are the surface atomic sites. In contrast to Fig. 4.7, there are no overhangs and voids that can created during growth.

Molecular-Beam Epitaxy was suggested to belong to have conservative growth dynamics. But recent studies [4.23, 4.24] argued that this may not be the case and at least the long time process is still governed by the non-conservative KPZ dynamics. However, metal thin film growth is indeed expected to proceed as a conservative growth. Overhangs and voids are unlikely to appear in the growth of metal thin films not only because surface diffusion plays a dominant role but also because metal atoms lack local chemical bonding such as dangling bonds. Even within the conservative growth models, recent theoretical study showed that there exists very rich universality classes which depend on different local hoping rules (different local diffusion mechnisms). [4.25] Also, a general cross-over behavior may occur from a short time behavior to a long time behavior with the value of β increased to 1. [4.26] Recent experiments in which self-affine fractals are characterized using different techniques indicate that the values of α measured for metal thin films ranges from 0.65 to 0.95. [4.9, 4.27 - 4.30]

In the last few years, there has been great effort in constructing theoretical models for thin film growth. Nevertheless, the authors believe that the field is still in its infancy and the fundamental understanding of the growth process is far from complete. We expect a substantial increase of research activities in this area in the near future.

REVIEW AND SUMMARY

Dynamic scaling hypothesis in thin film growth

It is proposed that the interface morphology during dynamic film growth is scale-invariant under the following scale transformation,

$$\begin{cases} \mathbf{r} \to \lambda \mathbf{r} & \text{(the lateral scale)} \\ z \to \lambda^{\alpha} z & \text{(the vertical scale)} \\ t \to \lambda^{\alpha/\beta} t & \text{(the time scale)} \end{cases} , \quad (4.3)$$

where α is the static exponent giving a measure of the short-range roughness and β is the dynamic exponent related to the global roughening evolution.

Scaling form of the interface width during dynamic growth

$$w(L, t) = L^{\alpha} g\left(\frac{t}{L^{\alpha/\beta}}\right), \quad (4.5)$$

with

$$g(x) \approx \begin{cases} x^{\beta} & \text{for } x << 1 , \\ \text{constant} & \text{for } x >> 1 , \end{cases} \quad (4.8)$$

where L is the lateral linear size of a section in the growing surface and w(L, t) is the interface width in the corresponding section. The w(L, t) is scale-invariant under the transformation of Eq. (4.3),

$$w(\lambda L, \lambda^{\alpha/\beta} t) = \lambda^{\alpha} w(L, t).$$

Scaling aspects during dynamic growth

(1) Global morphology (on the long-range scale, $L >> \xi$): According to Eqs. (4.5) and (4.8), the global interface structure undergoes both vertical roughening and lateral coarsening evolutions:

$$w(L, t) \sim t^{\beta} \text{ and } \xi(t) \sim t^{\beta/\alpha}.$$

The long-range surface morphology is scale-invariant in time (dynamic scaling),

$$\lambda^{\alpha} w(t_2) = w(t_1) \text{ and } \lambda \xi(t_2) = \xi(t_1),$$

where $\lambda = (t_1/t_2)^{\beta/\alpha}$.

(2) Local morphology (on the short-range scale, $L \ll \xi$): According to Eqs. (4.5) and (4.8), the local surface structure is a time-invariant and self-affine fractal,

$$w(L, t) \sim L^{\alpha}. \tag{4.9}$$

The short-range surface morphology is scale-invariant in space (static scaling),

$$w(\lambda L) = \lambda^{\alpha} w(L).$$

Dynamic scaling in the equal time height-height correlation function

$$H(\mathbf{r}, t) = \langle [z(\mathbf{r}, t) - z(0, t)]^2 \rangle$$
$$= 2[w(t)]^2 f\left(\frac{r}{\xi(t)}\right), \tag{4.18}$$

with

$$f(X) = \begin{cases} X^{2\alpha} & \text{for } X \ll 1, \\ 1 & \text{for } X \gg 1. \end{cases}$$

The equal time height-height correlation function is scale-invariant under the transformation of Eq. (4.3),

$$H(\lambda \mathbf{r}, \lambda^{\alpha/\beta} t) = \lambda^{2\alpha} H(\mathbf{r}, t).$$

The asymptotic expression can be derived from Eq. (4.18),

$$H(\mathbf{r}, t) \sim \begin{cases} r^{\alpha} & \text{for short ranges } L \ll \xi(t), \\ 2\, t^{2\beta} & \text{for long ranges } L \gg \xi(t). \end{cases}$$

Dynamic scaling in the height difference function

(1) Continuous surfaces
The height difference function is given by

$$C_c(k_\perp, \mathbf{r}, t) = e^{-[k_\perp w_c(t)]^2 f[\mathbf{r}/\xi(t)]}, \tag{4.20}$$

which is scale-invariant,

$$C_c(\lambda^{-\alpha} k_\perp, \lambda \mathbf{r}, \lambda^{\alpha/\beta} t) = C_c(k_\perp, \mathbf{r}, t).$$

(2) Crystalline surfaces

For diffraction conditions away from the "out-of-phase", $||[\phi]|| \neq \pi$, the height difference function is given by the lowest order approximation,

$$C_d(k_\perp, \mathbf{r}, t) \approx e^{-w_d(t)^2([\phi])^2 f[\mathbf{r}/\xi(t)]} . \qquad (4.23)$$

At the near out-of-phase condition, $||[\phi]|| \sim \pi$, due to the discrete lattice effect, one has to use the rigorous expression according to Eq. (2.22),

$$C_d(k_\perp, \mathbf{r}) = \frac{\sum_{n=-\infty}^{+\infty} e^{-w_d(t)^2 f[\mathbf{r}/\xi(t)]\,(\phi - 2\pi n)^2}}{\sum_{n=-\infty}^{+\infty} e^{-w_d(t)^2 f[\mathbf{r}/\xi(t)]\,(2\pi n)^2}} . \qquad (4.22)$$

$C_d(k_\perp, \mathbf{r}, t)$ is scale-invariant at the small $k_\perp$ condition, $\phi = k_\perp c \ll \pi$, but is not scale-invariant for large $k_\perp$, $\phi = k_\perp c \sim \pi$ or $\phi > \pi$. Such a scaling symmetry-breaking is due to the discrete lattice effect.

Growth Dynamics

Different growth processes would give different values of the growth exponents, α and β.

(1) Non-conservative growth process
Side growth is allowed with the creation of voids and overhangs and can be described by the KPZ equation. The volume during growth is non-conservative, i.e., the volume of the growing film is larger than that of the compact film of the same amount of deposited material. Examples are Eden model and the ballistic deposition.

(2) Conservative growth process
During the deposition, a compact film is created. The primary relaxation mechanism is surface diffusion. The mass and volume conservation laws play an important role in the growth since the desorption of atoms and formation of overhangs and voids are negligibly small.

The roughness exponent α for conservative growth processes has larger values than that for non-conservative growth processes.

REFERENCES

4.1 For review, see, F. Family, *Physica* **A168**, 561 (1990); *Dynamics of Fractal Surfaces*, ed. F. Family and T. Vicsek (World-Scientific, Singapore, 1990).

4.2 Also see, *On Growth and Form*, ed. H. E. Stanley and N. Ostrowsky (Martinus Nijhoff, Boston, 1986).

4.3 For recent reviews, see, *Solids Far From Equilibrium: Growth Morphology and Defects*, ed. C. Godriche (Cambridge University Press, New York, 1991), pp.432 and pp. 479.

4.4 *PHI Info*, No. 9202, June (Perkin Elmer Co., Minn., 1992).

4.5 H.-N. Yang, T.-M. Lu, and G.-C. Wang, *Phys. Rev.* **B47**, 3911 (1993).

4.6 H. E. Stanley in "Scaling phenomena in disordered systems" , *NATO ASI Series B: Physics* Vol. 133 (Plenum Press, New York, 1985), pp. 49.

4.7 R. Messier and J. E. Yehoda, *J. Appl. Phys.* **58**, 3739 (1985); J. E. Yehoda and R. Messier, *Appl. Surf. Sci.* **22/23**, 590 (1985).

4.8 B. B. Mandelbrot, *The Fractal Geometry of Nature* (Freeman, New York, 1982).

4.9 R. Chiarello, V. Panella, J. Krim, and C. Thompson, *Phys. Rev. Lett.* **67**, 3408 (1991).

4.10 H.-N. Yang, T.-M. Lu, and G.-C. Wang, *Phys. Rev. Lett.* **68**, 2612 (1992).

4.11 R. Jullien and R. Botet, *J. Phys.* **A18**, 2279 (1985).

4.12 J. G. Zabolitzky and D. Stauffer, *Phys. Rev. Lett.* **57**, 1809 (1986).

4.13 J. Kertész and D. E. Wolf, *J. Phys.* **A21**, 747 (1988).

4.14 D. E. Wolf and J. Kertész, *Europhys. Lett.* **4**, 651 (1987).

4.15 M. J. Vold, *J. Colloid Sci.* **14**, 168 (1959).

4.16 F. Family and T. Vicsek, *J. Phys.* **A18**, L75 (1985).

4.17 M. Kardar, G. Parisi, and Y. Zhang, *Phys. Rev. Lett.* **56**, 889 (1986).

4.18 J. Villain, *J. Phys.* I (France) **1**, 19 (1991).

4.19 D. E. Wolf and J. Villain, *Europhys. Lett.* **13**, 389 (1990).

4.20 Z.-W. Lai and S. Das Sarma, *Phys. Rev. Lett.* **66**, 2348 (1991).

4.21 L.-H. Tang and T. Nattermann, *Phys. Rev. Lett.* **66**, 2899 (1991).

4.22 A. Zangwill, C. N. Luse, D. D. Vvedensky, and M. R. Wilby, *Surf. Sci.* **274**, L529 (1992).

4.23 Hong Yan, *Phys. Rev. Lett.* **68**, 3048 (1992).

4.24 D. A. Kessler, H. Levine, and L. Sander, *Phys. Rev. Lett.* **69**, 100 (1992).

4.25 S. Das Sarma and S. V. Ghaisas, *Phys. Rev. Lett.* **69**, 3762 (1992).

4.26 J. G. Amar and F. Family, to be published.

4.27 M. W. Mitchell and D. A. Bonnell, *J. Mater. Res.* **5**, 2244 (1990).

4.28 J. M. Gómez-Rodríguez, A. M. Baró, and R. C. Salarezza, *J. Vac. Sci. Technol.* **B9**, 495 (1991).

4.29 Y.-L. He, H.-N. Yang, T.-M. Lu, and G.-C. Wang, *Phys. Rev. Lett.* **69**, 3770 (1992).

4.30 H.-N. Yang, A. Chan, and G.-C. Wang, *J. Appl. Phys.* **74** (1993), in press.

Chapter V DIFFRACTION STRUCTURE FACTOR FROM ROUGH GROWTH FRONTS

In the dynamic scaling description of growth, both the vertical correlation length (interface width) and the lateral correlation length are time dependent. In this chapter, we shall discuss in detail the characteristics of the diffraction structure factor from such a growth front. A particularly interesting result is the existence of a time-invariant structure at the "out-of-phase" diffraction conditions for a crystalline growth front. [5.1] It is a reflection of a time-invariant height-height correlation in the short-range regime in real space, i.e., the short-range roughness becomes stationary during growth. Methods of extracting various growth exponents from the structure factor at different diffraction conditions are given in detail. We also give an example of the experimental measurement of a particular epitaxial growth front using the HRLEED technique. [5.2]

§V.1 Scaling and the Discrete Lattice Effect on the Diffraction Structure Factor

According to the result given by Eq. (2.16), the diffraction structure factor from a dynamic growth front is the Fourier transform of the height difference function $C(k_\perp,\mathbf{r}, t)$,

$$S(\mathbf{k}_\parallel, k_\perp, t) = \int d^2r\, C(k_\perp, \mathbf{r}, t)\, e^{i\mathbf{k}_\parallel \cdot \mathbf{r}}. \tag{5.1}$$

The time-dependent height difference function, $C(k_\perp, \mathbf{r}, t) = \langle e^{ik_\perp[z(\mathbf{r}, t) - z(0, t)]} \rangle$, can be calculated by combining the equal time height-height correlation function, Eq.

(4.18), either with Eq. (2.19) for a continuous surface or with Eq. (2.22) for a crystalline surface.

For a continuous surface, since the height difference function $C_c(k_\perp, \mathbf{r}, t)$ given by Eq. (4.20) is a self-affine scale-invariant function, $C_c(\lambda^\alpha k_\perp, \lambda^{-1}\mathbf{r}, \lambda^{-\alpha/\beta}t) = C_c(k_\perp, \mathbf{r}, t)$, the corresponding diffraction structure factor, Eq. (5.1), must have a scaling relation,

$$S_c(\lambda\mathbf{k}_{\|}, \lambda^\alpha k_\perp, \lambda^{-\alpha/\beta}t) = \lambda^2 S_c(\mathbf{k}_{\|}, k_\perp, t).$$

However, as we have pointed out in §IV.2.3 for the height difference function, $C_d(k_\perp, \mathbf{r}, t)$, of a crystalline surface given by Eq. (4.22), the general self-affine scaling relation is broken by the "discrete lattice effect", i.e.,

$$C_d(\lambda^\alpha k_\perp, \lambda^{-1}\mathbf{r}, \lambda^{-\alpha/\beta}t) \neq C_d(k_\perp, \mathbf{r}, t).$$

Therefore, the diffraction structure factor, Eq. (5.1), for a crystalline surface usually does not obey the scaling relation of a continuous surface shown above, i.e.,

$$S_d(\lambda\mathbf{k}_{\|}, \lambda^\alpha k_\perp, \lambda^{-\alpha/\beta}t) \neq \lambda^2 S_d(\mathbf{k}_{\|}, k_\perp, t).$$

Only in the "small $k_\perp$" diffraction conditions, $|\phi| << \pi$, in which the height difference function is approximately self-affine scale-invariant, $C_d(\lambda^\alpha k_\perp, \lambda^{-1}\mathbf{r}, \lambda^{-\alpha/\beta}t) \approx C_d(k_\perp, \mathbf{r}, t)$, can the structure factor have an approximate scaling relation,

$$S_d(\lambda\mathbf{k}_{\|}, \lambda^\alpha k_\perp, \lambda^{-\alpha/\beta}t) \approx \lambda^2 S_d(\mathbf{k}_{\|}, k_\perp, t), \qquad (|\phi| << \pi).$$

This scaling relationship is identical to that from a continuous surface. The reason for this is the following: the "small $k_\perp$" diffraction condition gives a diffraction discrete lattice aspect can thus be ignored and the surface looks like a continuous one. However, we must point out that the lack of the scaling behavior at large ϕ in $S_d(\mathbf{k}_{\|},k_{\perp},t)$ for the crystalline surface does not prevent us from extracting the roughness parameter associated with the height-height correlation function $H_d(\mathbf{r},t)$.

The diffraction structure factor from a dynamic growth front can be divided into two parts: one is a sharp central δ-function and another one contains a broad diffuse component,

$$S(\,\mathbf{k}_{\|},k_{\perp},t\,) = \int d^2r\, C_{\infty}(k_{\perp},t)\, e^{\,i\mathbf{k}_{\|}\cdot\mathbf{r}} + \int d^2r\, \Delta C(\,k_{\perp},\mathbf{r},t\,)\, e^{\,i\mathbf{k}_{\|}\cdot\mathbf{r}}$$

$$= (2\pi)^2\, C_{\infty}(k_{\perp},t)\, \delta(\,\mathbf{k}_{\|}\,) + S_{diff}(\,\mathbf{k}_{\|},k_{\perp},t\,)\,. \qquad (5.1')$$

Here the diffuse structure factor $S_{diff}(\mathbf{k}_{\|},k_{\perp},t)$ is given by

$$S_{diff}(\,\mathbf{k}_{\|},k_{\perp},t\,) = \int d^2r\, \Delta C(\,k_{\perp},\mathbf{r},t\,)\, e^{\,i\mathbf{k}_{\|}\cdot\mathbf{r}}\,, \qquad (5.2)$$

with $C_{\infty}(k_{\perp},t) = C(k_{\perp},\mathbf{r}\to\infty,t)$ and $\Delta C(k_{\perp},\mathbf{r},t) = C(k_{\perp},\mathbf{r},t) - C(k_{\perp},\mathbf{r}\to\infty,t)$. (See Chapter III for a similar treatment of the static case.) Since the height-height correlation function for the dynamic growth front, as shown in Eq. (4.18), has the same form as that for the case of the static rough surface, as given by Eq. (2.27), the conclusions for the diffraction problem must be similar in these two cases. For the present dynamic growth problem, we can therefore utilize the basic results obtained from the static rough surface, as derived in Chapter III.

§V.2 The Time-invariant Structure Factor

In Chapter III, we showed that the diffraction structure factor has an asymptotic form under the condition of $\Omega >> 1$, given by Eq. (3.20). Accordingly, for the present dynamic growth interface, if $\Omega >> 1$, we have

$$S(\mathbf{k}_{\|}, k_{\perp}, t) \approx \begin{cases} 2\pi\,(\eta k_{\perp}^{-\frac{1}{\alpha}})^2\, F_{\alpha}(k_{\|}\eta k_{\perp}^{-\frac{1}{\alpha}}) & \text{for a continuous surface ,} \\ 2\pi\,(\eta[\phi]^{-\frac{1}{\alpha}})^2\, F_{\alpha}(k_{\|}\eta[\phi]^{-\frac{1}{\alpha}}) & \text{for a crystalline surface .} \end{cases} \quad (5.3)$$

As we have discussed in §IV.2.1, $\eta = \xi(t)w(t)^{-\frac{1}{\alpha}}$ is a time-independent quantity, which, in the case of a crystalline surface, is interpreted as proportional to the average terrace size during growth. Thus, $S(\mathbf{k}_{\|}, k_{\perp}, t)$ at $\Omega >> 1$, given by Eq. (5.3), is a time-invariant structure factor determined by the short-range and time-independent parameters, α and η.

As shown in Fig. 4.5(a), the equal time height-height correlation is time-invariant in the short-range regime, $r << \xi(t)$. The height difference function $C(k_{\perp}, \mathbf{r}, t)$ at $\Omega >> 1$ is thus dominated by the short-range and time-invariant characteristics of the growing surface. The corresponding diffraction structure factor, which is the Fourier transform of $C(k_{\perp}, \mathbf{r}, t)$, must become time-independent. In §III.2, we obtained an analytic expression for the short-range height difference function, Eq. (3.18), which is also correct in the present growth problem. That is, for $\Omega >> 1$, we have

$$C(k_{\perp}, \mathbf{r}, t) \approx e^{-(\Omega/\xi^{2\alpha})\, r^{2\alpha}} = \begin{cases} e^{-k_{\perp}^2 (r/\eta)^{2\alpha}} & \text{for a continuous surface ,} \\ e^{-[\phi]^2 (r/\eta)^{2\alpha}} & \text{for a crystalline surface .} \end{cases} \quad (5.4)$$

This is obtained by inserting the short-range and time-invariant height-height correlation function, $H(\mathbf{r}, t) \approx 2\left(\frac{r}{\eta}\right)^{2\alpha}$, into either Eq. (2.19) or Eq. (2.22). If the diffraction structure factor Eq. (5.1) is calculated using the time-invariant height difference function, Eq. (5.4), we can immediately obtain Eq. (5.3).

We remind the readers that Eq. (5.3) does not apply to the case in a crystalline surface at the vicinity of the out-of-phase diffraction conditions due to the discrete lattice effect, as discussed in §III.4. The rigorous result at the near out-of-phase conditions can be calculated numerically by combining Eq. (4.22) with Eq. (5.1). The modification due to the discrete lattice effect is similar to that for a static rough surface demonstrated in Fig. 3.6. However, as we have pointed out in §III.4, such a modification should not change the basic conclusion that at $\Omega >> 1$, the diffraction structure factor is time-invariant.

In conclusion, the time-invariant diffraction behavior occurs under the condition $\Omega >> 1$, at which, the time-dependent central δ-peak drops sharply to zero while the remaining diffuse component is purely a contribution from the short-range scattering events.

According to Eq. (3.8), the crucial parameter $\Omega = \Omega(t)$ in the present growth problem is given by

$$\Omega(t) \equiv \begin{cases} k_\perp^2 w_c(t)^2 & \text{for a continuous surface ,} \\ [\phi]^2 w_d(t)^2 & \text{for a crystalline surface .} \end{cases} \qquad (5.5)$$

The condition $\Omega(t) >> 1$ gives a restriction for the vertical wavevector, $k_\perp$. One may define, in the reciprocal space of $k_\perp$, the time-invariant zones in which $\Omega(t) >> 1$. The diffraction at this condition is dominated only by the short-range behavior as characterized by α and η. This treatment enables us to separate the time-invariant part from the time-dependent part in a diffraction experiment, so that the scaling properties can be measured and interpreted readily. For a continuous surface, the time-invariant zone corresponding to the larger $k_\perp$ diffraction condition is shown as the shaded area in Fig. 5.1(a). For a crystalline surface, shown in Fig. 5.1(b), these time-invariant zones are centered at the out-of-phase conditions, $k_\perp = \phi/c = (2n - 1)\pi/c$, with the range determined by $\Omega(t) = [\phi]^2[w_d(t)]^2 >> 1$.

The exponent α and the average terrace size η can be measured experimentally according to the methods discussed in §III.3.3. We must be aware of the discrete lattice effect for the case of a crystalline surface, where a more reliable method to extract α should come from the rigorous relation of FWHM vs. ϕ shown in Fig. 3.10.

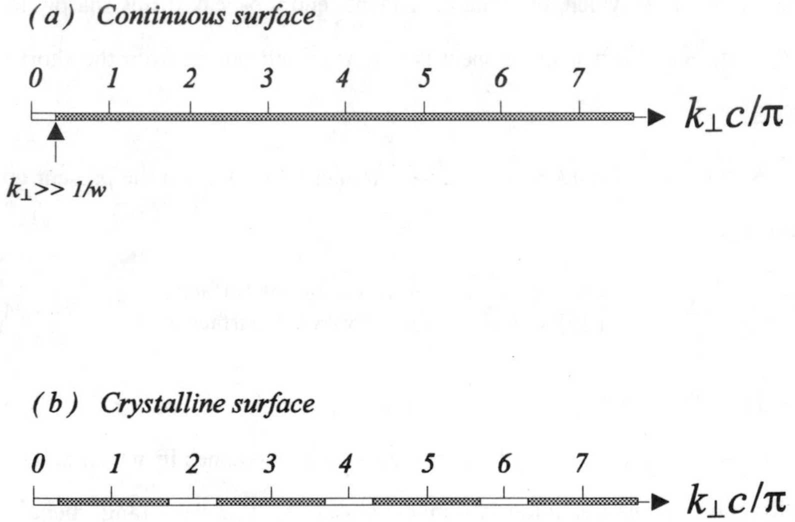

Fig. 5.1 Schematics of time-invariant zones (shaded area) in the reciprocal space of $k_\perp$. In order to compare (a) with (b), we use the same units, π/c, for $k_\perp$.

§V.3 The Time-dependent Structure Factor

Away from the time-invariant zone with $\Omega(t) << 1$, where $\Omega(t)$ is defined in Eq. (5.5), the diffraction structure factor is sensitive to the global (long-range) surface properties such as the time-dependent evolution of the growing interface. In contrast to the condition, $\Omega(t) >> 1$, which defines the time-invariant zone with a diffraction structure factor sensitive only to the short-range behavior, the condition, $\Omega(t) << 1$, can be used to define a time-dependent zone with a structure factor which is particularly sensitive to the long-range behavior. $\Omega(t) << 1$ implies the small $k_\perp$ diffraction condition for a continuous surface and the near in-phase diffraction condition, $[\phi] \sim 0$, for a crystalline surface.

Under the condition, $\Omega(t) << 1$, the diffraction structure factor from the growth front, Eq. (5.1'), can be simplified according to Eq. (3.13) as

$$S(\mathbf{k}_{\|}, k_\perp, t) \approx (2\pi)^2 e^{-\Omega(t)}\delta(\mathbf{k}_{\|}) +$$

$$+ 2\pi\Omega(t)\xi^2(t)e^{-\Omega(t)} \int_0^\infty XdX[1-f(X)]J_0[Xk_{\|}\xi(t)]. \qquad (5.6)$$

Equation (5.6) indicates that for $\Omega(t) << 1$, both the δ–component and the diffuse component are time-dependent. The δ–peak intensity, $I_\delta \propto e^{-\Omega(t)}$, is a function of the interface width w(t) through Eq. (5.5). The time-dependent interface width w(t) can be measured according to the methods introduced in §III.3.1. Since $w(t) \sim t^\beta$, we have $Ln(I_\delta) \propto -t^{2\beta}$ according to Eq. (5.5). The growth exponent β can then be obtained from the measured δ–peak intensity as a function of the growth time, t.

On the other hand, the diffuse line shape has a form,

$S_{diff}(\mathbf{k}_{\|}, k_{\perp}, t) \approx$

$$\approx 2\pi\Omega(t)\xi^2(t)\, e^{-\Omega(t)} \int_0^{\infty} XdX[1-f(X)]J_0[Xk_{\|}\xi(t)], \quad (\Omega(t) << 1), \quad (5.7)$$

which is a function of the interface width w(t) and the lateral correlation length $\xi(t)$. As shown in Eq. (3.14), the FWHM of the diffuse line shape of Eq. (5.7) is given by

$$\text{FWHM} = \frac{2Y_g}{\xi(t)},$$

where the constant Y_g satisfies the following equation,

$$\int_0^{\infty} XdX[1-f(X)]J_0(XY_g) = 0.5\int_0^{\infty} XdX[1-f(X)].$$

The FWHM of the diffuse line shape at $\Omega(t) << 1$ is thus inversely proportional to the time-dependent lateral correlation length, $\xi(t)$. This relationship can be utilized to determine $\xi(t)$. Since $\xi(t) \sim t^{\beta/\alpha} = t^{1/z}$, we have

$$\text{FWHM} \sim t^{-\beta/\alpha} = t^{-1/z}. \quad (5.8)$$

The ratio of the exponents $z = \alpha/\beta$ can thus be obtained from the measured FWHM as a function of growth time.

In conclusion, at $\Omega(t) << 1$, corresponding to either the small $k_{\perp}$ diffraction condition or the near in-phase condition, both the δ–intensity and the FWHM of the diffuse line shape continuously decay with time, i.e., $Ln(I_{\delta}) \propto -t^{2\beta}$ and FWHM ~

$t^{-\beta/\alpha} = t^{-1/z}$, respectively. This time-dependent behavior exhibits continuous global roughening and coarsening evolution during the dynamic growth process.

§V.4 An Experimental Measurement of An Epitaxial Growth Front

In this section we present an experimental measurement of a particular epitaxial growth front, namely, the molecular beam epitaxial growth of Fe on Fe(100) surface using the HRLEED technique.[5.2] A clear indication of the existence of a time-invariant structure factor is demonstrated. The time-invariant structure factor implies a stationary density of steps. The roughness parameter is also extracted from the energy dependence of the structure factor measured near the out-of-phase diffraction condition.

In our experiment, a buffer layer consisting of about 100 atomic layers of Fe is first formed on the Au(001) surface to release the interface strain. The Fe film has been shown to have a crystalline structure. The (00) beam angular distribution of HRLEED intensity showed a very narrow profile sitting on a diffuse intensity profile at the Fe-Fe out-of-phase diffraction condition. At the in-phase diffraction condition, the diffuse intensity disappeared and the profile resembled the instrument response of the HRLEED system. We shall see later that this surface is basically flat and contained steps confined to only the top two layers of atoms. The interface width is negligibly small. We then begin to study the dynamic scaling growth with a deposition rate ~2 monolayer/min. The surface morphology was characterized by HRLEED after every 3-minute deposition interval. No significant self-annealing effect can be observed after each Fe deposition for a very long time. The diffraction line shapes remain the same even after one or two days.

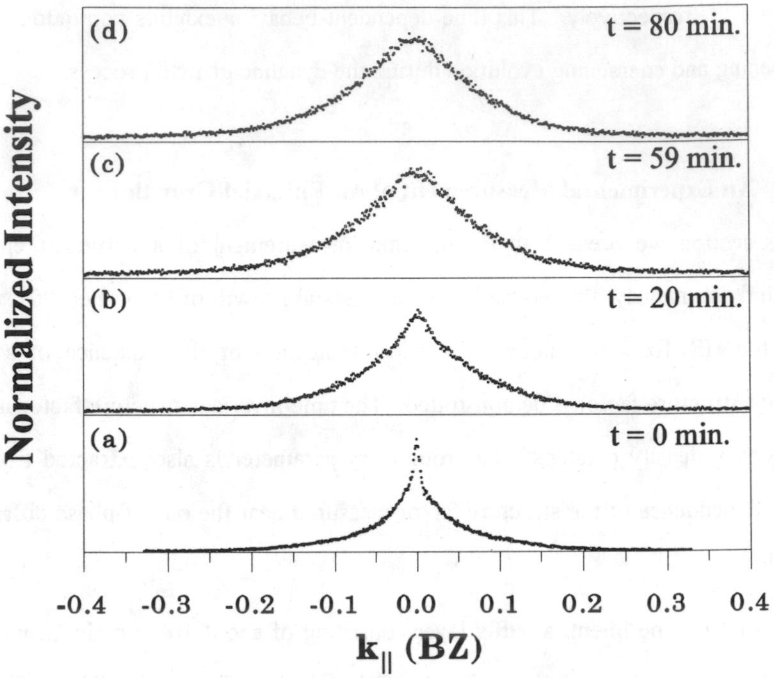

Fig. 5.2 Line shapes of the (00) HRLEED beam intensity at E = 40.0 eV, the out-of-phase diffraction condition, $\phi \approx 3\pi$, at different times during growth. For Fe(001), the dimension of the Brillouin zone is BZ = 2.1893 $Å^{-1}$. Shown in Fig. 5.2(a), the central part of the line shape has a Gaussian-like portion with a narrow width equal to the instrument response of HRLEED. This central portion corresponds to the δ-component of the line shape.

§V.4.1 *Time-dependent diffraction line shape during growth process*

The (00) beam angular profiles of the HRLEED intensity were measured as a function of the incident electron beam energy, E, after each deposition. Figure 5.2(a) shows the line shapes of the (00) beam intensity measured at different growth times at E = 40.0 eV, corresponding to the Fe-Fe out-of-phase diffraction condition: $\phi \approx$

3π. In Fig. 5.2(a), the line shape at t = 0 is basically a δ-like central spike (convoluted with the narrow instrument response) superimposed on a broad diffuse profile. As the film grows thicker and thicker, the central spike intensity quickly drops to zero, as shown in Figs. 5.2(b), 5.2(c) and 5.2(d). All these curves are one-dimensional scans of the two-dimensional angular profiles.

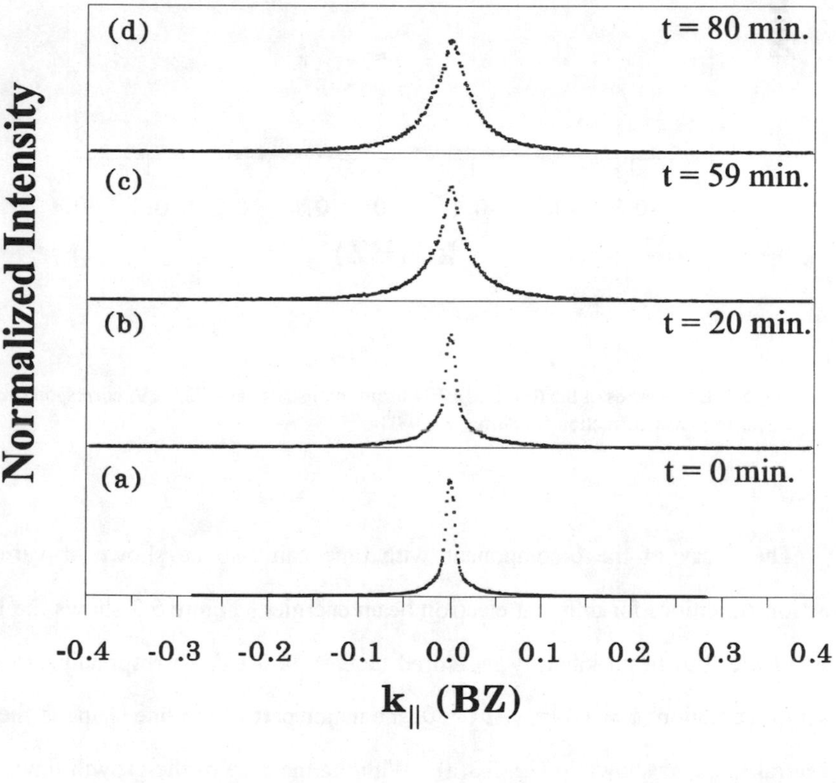

Fig. 5.3 Growth-time-dependent Line shapes of the (00) HRLEED beam intensity at E = 54.0 eV, corresponding to the diffraction condition, $\phi \approx 3.44\pi$.

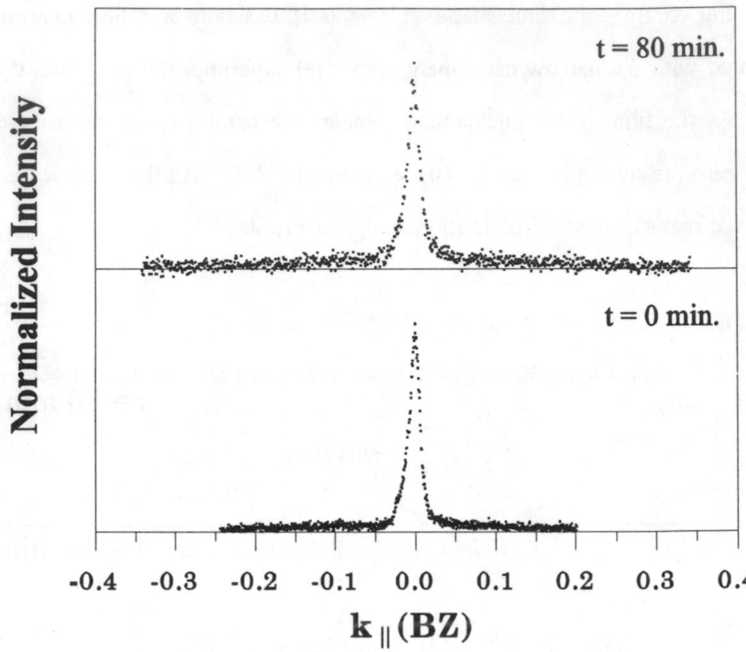

Fig. 5.4 Line shapes of the (00) HRLEED beam intensity at E = 72.0 eV, corresponding to a near in-phase diffraction condition, $\phi \approx 4.03\pi$.

The decay of the δ-component with time can also be shown at various diffraction conditions for different electron beam energies. Figure 5.3 shows the line shapes of the (00) beam intensity measured at E = 54.0 eV, corresponding to the diffraction condition: $\phi \approx 3.44\pi$. At t = 0, the major part of the line shape is the δ-like central spike, as shown in Fig. 5.3(a). With the increase of the growth time, the diffuse component appears and becomes more and more significant. However, unlike the case at the out-of-phase condition, $\phi \approx 3\pi$, as shown in Figs. 5.2(c) and 5.2(d), the δ-like central spike which decays with time does not disappear as quickly at this diffraction condition, $\phi \approx 3.44\pi$, as shown in Figs. 5.3(c) and 5.3(d). In Fig. 5.4, we also plot the line shapes of the (00) beam intensity measured at E = 72.0 eV, corresponding to a near in-phase condition: $\phi \approx 4.03\ \pi$. We note that this near in-phase line shape is basically a δ-component with a small background noise. Besides, it shows a very small change during growth even though the line shape at the out-of-phase condition changes quite dramatically.

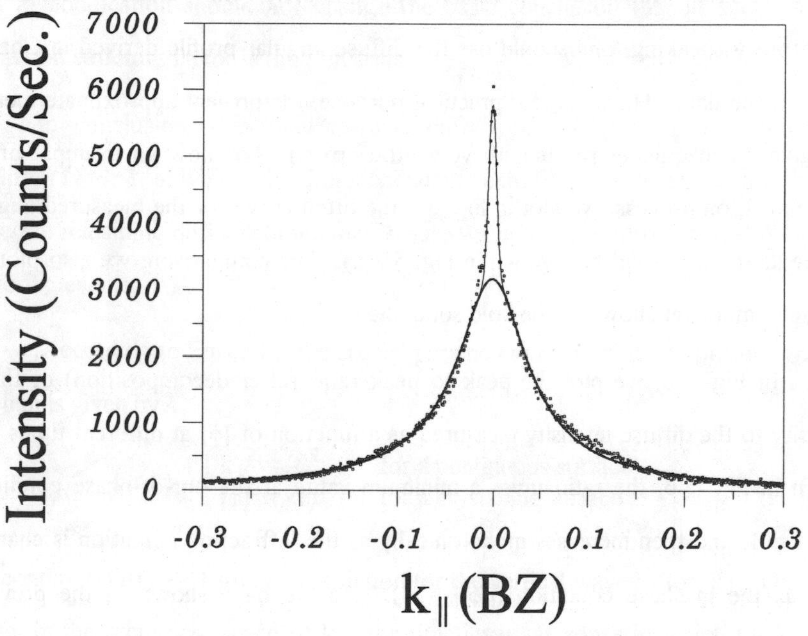

Fig. 5.5 The fitted curve (thin solid line) for the angular profile shown in Fig. 5.2(a). The bold solid line shows the diffuse component in the line shape.

In order to describe quantitatively the time dependent behavior of the line shapes at different diffraction conditions shown in Figs. 5.2, 5.3 and 5.4, we need to calculate the peak to peak ratio of the δ-intensity, I_δ, to the diffuse intensity, I_{diff}. The ratio can be obtained after decomposing the profiles into the δ–component and the diffuse component. During the decomposition, we fit the diffraction profiles using a sum of a sharp Gaussian function and a broad 1D Lorenztian function. The Gaussian function represents the δ–component convoluted with the instrumental response (a Gaussian-like shape in our HRLEED system, as shown in Fig. 5.4). The 1D Lorenztian function represents approximately the diffuse component. (Rigorously speaking, one should use the diffuse angular profile derived in Chapter IV to fit the data. However, for practical purpose, the present approximate method can give much quicker result with very little error.) To show an example of the decomposition process, we plot in Fig. 5.5 the fitted curve for the measured angular profile as the thin solid line shown in Fig. 5.2(a). For comparison, we also plot the diffuse component shown as the bold solid line.

In Fig. 5.6, we plot the peak to peak ratio (after decomposition) of the δ-intensity to the diffuse intensity measured as a function of **[ϕ]** at different times. As shown in Fig. 5.6, the ratio gives a minimum value at the out-of-phase condition, $|[\phi]|/\pi = 1$, and then increases monotonically as the diffraction condition is changed towards the in-phase condition ($|[\phi]| = 0$). As we have shown in the previous section, the δ-intensity is a measure of long-range surface flatness during growth. The δ–peak intensity, as shown in Eq. (5.1'), is given by

$$I_\delta \propto C_\infty(k_\perp, t) = C_d(k_\perp, \mathbf{r}\to\infty, t) \sim e^{-[\phi]^2 w_d(t)^2}$$

which is very sensitive to the interface width $w_d(t)$. Besides, I_δ is also sensitive to the diffraction condition $[\phi]$. The δ–intensity has a minimum value at the out-of-phase condition, $[\phi] = \pi$, and increases monotonically as $[\phi]$ changes from $[\phi] = \pi$ to the in-phase condition, $[\phi] = 0$. The decay of the δ-intensity with $|[\phi]|$ should be much quicker when $[\phi]$ is close to the out-of-phase condition. All these aspects are qualitatively consistent with the peak to peak ratio of the δ-intensity to the diffuse intensity measured as a function of $[\phi]$, shown in Fig. 5.6.

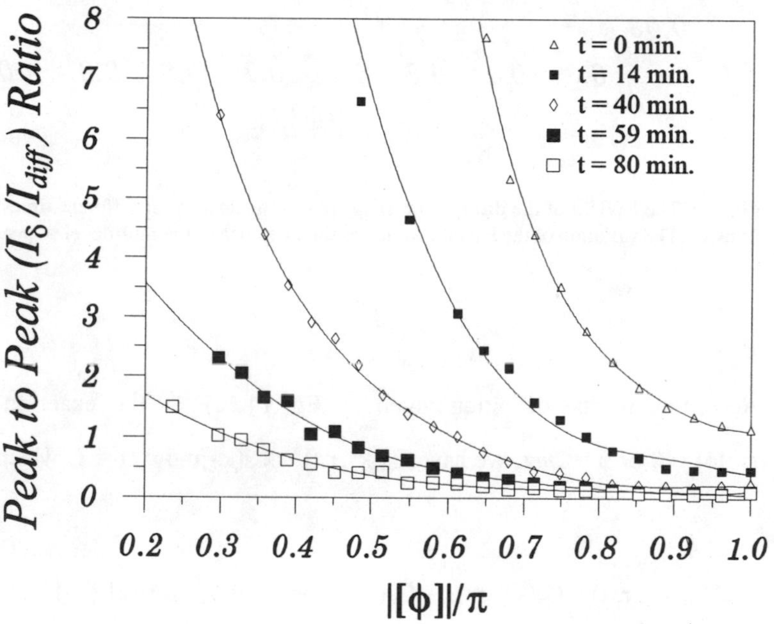

Fig. 5.6 The peak to peak ratio of the δ-intensity to the diffuse intensity is plotted as a function of $|[\phi]|$ at different times during growth. The solid curves are guidelines. The data has an uncertain $\sim \pm 10\%$.

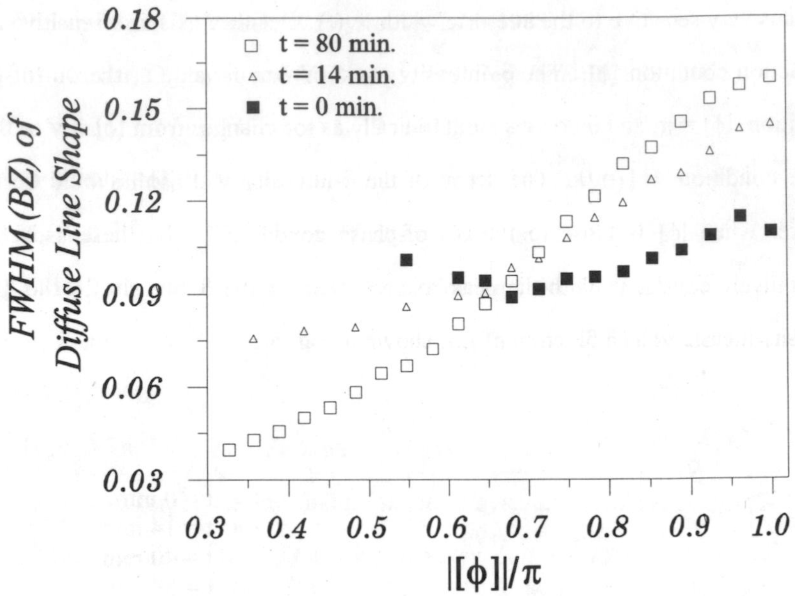

Fig. 5.7 The FWHM of the diffuse line shape plotted against [ϕ] at different times during growth. The variation of the FWHM indicates the existence of the multilevel step surface.

According to the definition given by Eq. (4.22), at the exact in-phase condition, $[\phi] = 0$ or $\phi = 2m\pi$, we have $C_d(k_\perp,\mathbf{r},t) = C_d(k_\perp,\mathbf{r}\to\infty,t) = 1$, which leads to

$$\Delta C_d(k_\perp, \mathbf{r}, t) \equiv C_d(k_\perp, \mathbf{r}, t) - C_d(k_\perp, \mathbf{r}\to\infty, t) = 0, \quad (\text{for } [\phi] = 0).$$

Accordingly, the diffuse component, $S_{diff}(\mathbf{k}_\parallel, k_\perp, t) = \int d^2r\, \Delta C(k_\perp, \mathbf{r}, t)\, e^{i\mathbf{k}_\parallel \cdot \mathbf{r}} = 0$. This is consistent with the in-phase line shape shown in Fig. 5.4. The disappearance of the diffuse line shape is due to the constructive interference of the diffraction from the different levels of the crystalline surface at the in-phase condition. In other words, at the in-phase condition, the multilevel stepped surface looks like a perfect flat surface from the diffraction point of view. Therefore, the electron wave at the in-phase diffraction condition is not sensitive to surface steps and the line shape should remain sharp (equal to the instrument response) at all times during growth, as shown in Fig. 5.4.

Since $I_\delta \sim e^{-[\phi]^2 w_d(t)^2}$, the continuous decrease of the δ-intensity with time, as shown in Figs. 5.2, 5.3 and 5.6, is a clear indication of the increase of interface width, $w_d(t)$. Such a growth-induced multilevel roughening can also be seen in the evolution of the energy-dependent diffuse intensity distribution. Figure 5.7 is the plot of the full-width-at-half-maximum (FWHM) of the diffuse line shape as a function of [ϕ] measured at different times. At t=0, the plot shows only a slight variation, implying that the surface fluctuation is confined nearly to the top two levels (the restricted two-level system would give a constant curve of FWHM vs. [ϕ] [5.3, 5.4]). At t > 0, as shown in Fig. 5.7, the growth leads to a significant variation of the FWHM (shown as open triangles and open squares) as a function of [ϕ] and therefore, the formation of a multilevel step structure.

§V.4.2 *Time-invariant diffraction line shape at the out-of-phase condition*

In contrast to the time-dependent evolution shown above, at the out-of-phase condition, where the δ-intensity decays sharply to zero, the diffraction is not sensitive to the long-range change during growth. As shown at 59 min. and 80 min. in Figs. 5.2(c) and 5.2(d), the residual diffuse angular profiles do not undergo any further change after the disappearance of the δ-component. Such a time-invariant behavior can be demonstrated more clearly in Fig. 5.8 where the FWHM of the entire angular profile at the out-of-phase condition is plotted as a function of time. After the gradual increase due to the initial roughening evolution, as shown in the first 40 minutes, the entire profile remains stationary and the FWHM reaches a constant value for t > 40 min. (At t=107 min., about 200 atomic layers of Fe have grown). We have shown in §V.2 that the time-invariant diffraction structure factor occurs when $w_d(t)^2[\phi]^2 >> 1$. With the increase of the interface width, time-invariance should therefore appear first at the out-of-phase condition where $||[\phi]||$ has a maximum value, $||[\phi]|| \sim \pi$. Our measurements confirms the existence of the short-range time-invariant characteristic as predicted in §V.2. The observed time-invariant behavior at this diffraction condition is therefore an indication that the system has already reached a dynamic scaling regime.

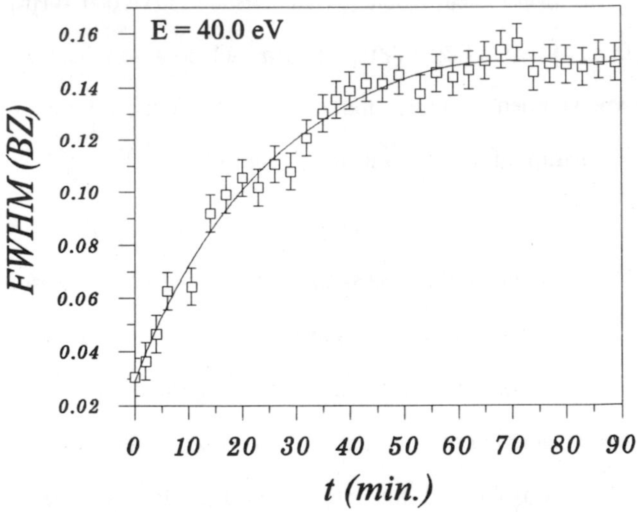

Fig. 5.8 The time-dependent FWHM of the diffraction line shape, shown as open squares, is measured at $\phi = 3\pi$ (E = 40.0 eV), the out-of-phase condition, where the solid curve is a guide to the eyes.

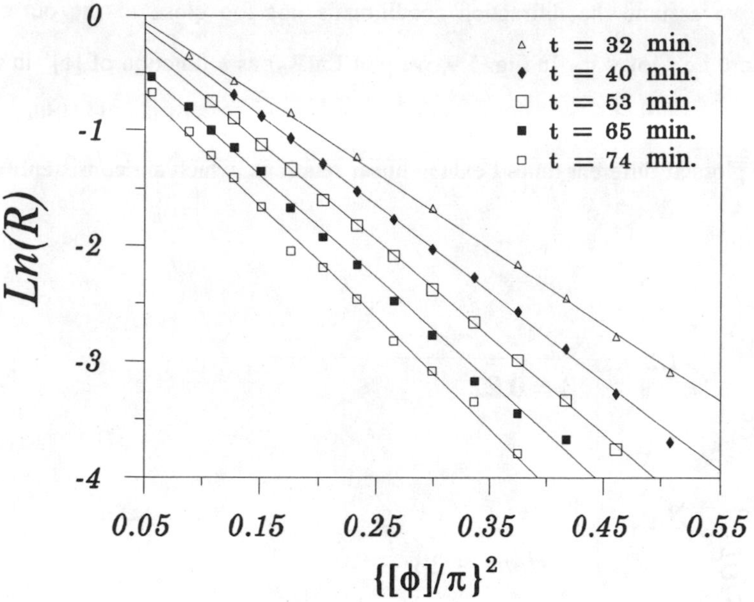

Fig. 5.9 LnR, (the natural logarithm of the ratio R of the integrated δ-intensity to the total integrated intensity), is plotted as a function of $\{[\phi]/\pi\}^2$ at different times during growth. The plot should give a slope, $-[\pi w_d(t)]^2$, according to Eq. (5.9).

§V.4.3 *Quantitative measurement of the growth exponents*

In order to obtain the interface width $w_d(t)$, we measure the integrated intensities for both the δ-component and the diffuse profile. According to the derivation shown in §III.3.1, the ratio, R_δ, of the integrated δ–intensity to the total integrated diffraction intensity has a simple relation,

$$R_\delta = \frac{\int d^2k_{\|}\, I_\delta(\mathbf{k}_{\|}, k_\perp, t)}{\int d^2k_{\|}\, I(\mathbf{k}_{\|}, k_\perp, t)} = C_d(k_\perp, \mathbf{r}\to\infty, t) \sim e^{-[\phi]^2 w_d(t)^2}, \qquad (5.9)$$

where we assume the diffraction condition is not too close to the out-of-phase condition, i.e., $[\phi] < \pi$. In Fig. 5.9, we plot $Ln(R_\delta)$ as a function of $[\phi]^2$ in units of π^2. The integration of the diffraction intensity is over 50% of the 2D Brillouin zone. All the plots at different times t exhibit linear relations, which are consistent with Eq. (5.9).

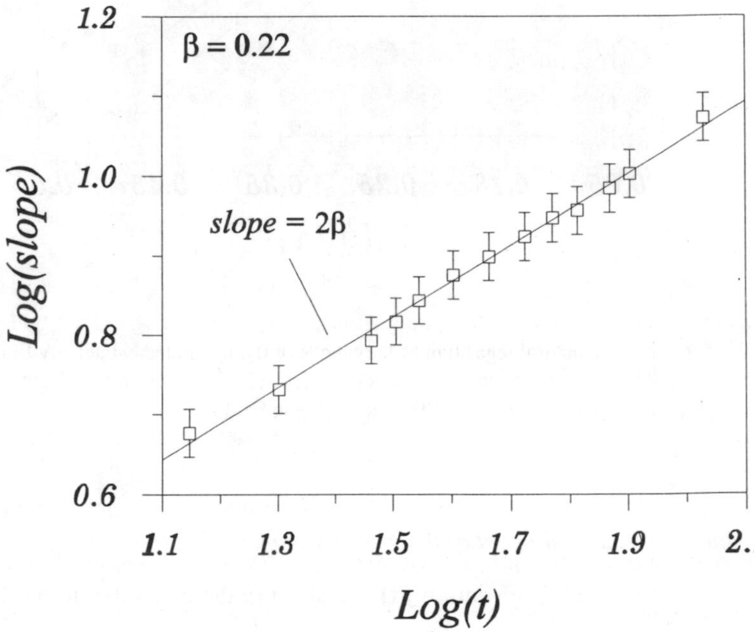

Fig. 5.10 The slopes measured in Fig. 5.9 plotted as a function of time on the log-log scale. The linear relation in this plot indicates that the interface width grows in a form of power-law, $[w_d(t)]^2 \propto t^{2\beta}$, with $\beta = 0.22$.

The plot of $Ln(R_\delta)$ vs. $([\phi]/\pi)^2$ should give a slope that is equal to $-[\pi w_d(t)]^2$, as shown in Eq. (5.9). The systematic increase of the slope with time in Fig. 5.9 is therefore another indication of the growth-induced roughening evolution. Shown in Fig. 5.10 is the log-log plot of the slope measured from Fig. 5.9 as a function of the growth time, i.e., Log(|slope|) vs. Log(t). The plot in Fig. 5.10 seemingly demonstrates a linear relation, i.e.,

$$\text{Log}(|\text{slope}|) = \gamma_o \text{Log}(t) + \text{constant},$$

where γ_o is a constant that can be obtained from the slope of the straight line shown in Fig. 5.10. This linear relation therefore leads to a power-law form,

$$|\text{slope}| \propto t^{\gamma_o}.$$

On the other hand, according to Eq. (5.9) and Eq. (4.12), $|\text{slope}| = [\pi w_d(t)]^2 \propto t^{2\beta}$, which indicates that $\gamma_o \equiv 2\beta$. The growth exponent β can then be extracted from Fig. 5.10 and is found to be $\beta = 0.22 \pm 0.02$.

The short-range self-affine property of the growing interface can be further examined from the diffuse intensity distribution measured at different diffraction conditions. The diffuse line shape can be fitted with the formula given by combining Eq. (4.22) with Eq. (5.1), where one can use a phenomenological scaling function,[5.5] $f(X) = 1 - e^{-X^{2\alpha}}$, as shown in Eq. (2.28). We found that the diffuse line shape of I_{diff} vs. $\mathbf{k}_{||}$ can be fitted with a range of α from 0.5 to 1.0. The line shape of I_{diff} vs. $\mathbf{k}_{||}$ may not be very sensitive to the variation of α because the FWHM of the diffuse profile is mainly determined by the lateral property such as η or ξ. The insensitivity may also be attributed to the uncertainty caused by the background noise in the measured diffuse intensity. However, the plot of FWHM vs. [ϕ] should be quite sensitive to α because this relation depends on both the vertical and the lateral correlations during growth. We have shown in §V.2 and §V.3 that the FWHM of the diffuse line shape can vary from the time-invariant value $\propto [\phi]^{\frac{1}{\alpha}}\eta^{-1}$ at the vicinity of the out-of-phase condition to the time-dependent value $\propto \xi(t)^{-1} = \eta^{-1}w(t)^{-\frac{1}{\alpha}}$, at the near in-phase condition. The FWHM oscillates between the maximum at the out-of-phase condition and the minimum at the near in-phase condition.

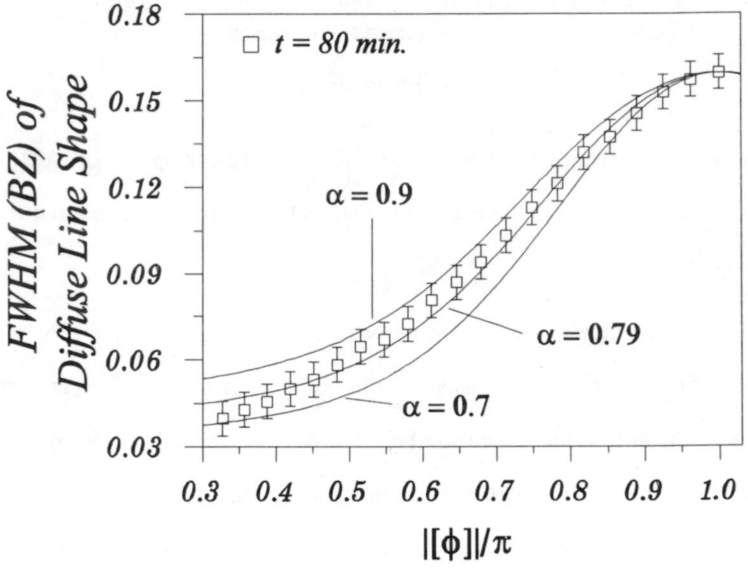

Fig. 5.11 The extraction of α from the variation of the FWHM of the diffuse line shape vs. ||[ϕ]||. The solid curve is the fit of the experimental data (open squares) taken at t = 80 min., corresponding to α = 0.79. To show the sensitivity of the fit, we also plot two simulated curves which correspond to α = 0.7 and α = 0.9.

The value of α can be extracted from the measured FWHM vs. $[\phi]$ plot in the time-invariant out-of-phase regime, where the relation is only determined by α and η as $w(t)^2[\phi]^2 >> 1$. However, as we have pointed out before (see §III.4), the sensitivity to α can be significantly reduced at the out-of-phase regime due to the rounding of the FWHM vs. $[\phi]$ curve caused by the discrete lattice effect. In order to obtain a more accurate value of α, we would rather extend the fit from the out-of-phase regime to the time-dependent regime down to $\|[\phi]\| \sim 0.3\pi$. During the fit, α and η are adjustable parameters while the time-dependent $w_d(t)$ is treated as a known parameter that varies from 0.8 to 1.1 obtained from Fig. 5.9. The solid curve in Fig. 5.11 shows the best fit for the experimental data, which gives, $\alpha = 0.79 \pm 0.05$ and $\eta = 14.0 \pm 2.0$ Å.

Our present experiment suggests that the deposition of Fe on Fe(001) belongs to a conservative growth process in which the theoretical studies have predicted that the scaling exponents are $\alpha = 2/3$ and $\beta = 1/5$ for a two-dimensional surface. As we have pointed out in §IV.3, overhangs and voids are unlikely to appear in metal films and the growth of a metal film normally proceeds as a conservative process. The exponents, $\alpha = 0.79 \pm 0.05$ and $\beta = 0.22 \pm 0.02$, obtained from the metal film of Fe on Fe(001) in the present experiment, are thus considered to be different from the prediction of non-conservative dynamics, but agree reasonably well with conservative growth models.

Thus far, very few experiments [5.6 - 5.9] have been designed to study the details of the dynamic scaling growth processes. The present experiment [5.2] represents such an effort. Both the time-invariant and time-dependent characteristics have been observed during growth. Quantitative measurements of the growth exponents suggest that the growth of this metal thin film belongs to a universality class predicted by conservative growth models.

REVIEW AND SUMMARY

Dynamic scaling and the discrete lattice effect on diffraction structure factor

(1) Continuous surfaces

The diffraction structure factor is scale-invariant,

$$S_c(\lambda \mathbf{k}_{\|}, \lambda^{\alpha} k_{\perp}, \lambda^{-\alpha/\beta} t) = \lambda^2 S_c(\mathbf{k}_{\|}, k_{\perp}, t),$$

because of the scale-invariant height difference function, $C_c(\lambda^{-\alpha} k_{\perp}, \lambda \mathbf{r}, \lambda^{\alpha/\beta} t) = C_c(k_{\perp}, \mathbf{r}, t)$.

(2) Crystalline surfaces

The diffraction structure factor is only scale-invariant at the small $k_{\perp}$, $\phi = k_{\perp} c << \pi$. $S_d(\mathbf{k}_{||}, k_{\perp}, t)$ is not scale-invariant for large $k_{\perp}$ condition, $\phi = k_{\perp} c \sim \pi$ and $\phi > \pi$, in which the scaling symmetry-breaking occurs due to the discrete lattice effect.

The long-range & time-dependent structure factor during dynamic growth

For $\Omega(t) << 1$, the diffraction structure factor $S(\mathbf{k}, t)$ is time-dependent and is determined by the long-range parameters, $w(t) \sim t^{\beta}$ and $\xi(t) \sim t^{\beta/\alpha}$,

$$S(\mathbf{k}, t) \approx (2\pi)^2 e^{-\Omega} \delta(\mathbf{k}_{\|}) + 2\pi \Omega \xi^2 e^{-\Omega} F[k_{\|} \xi(t)],$$

where $F(Y) = \int_0^{\infty} X dX \, [1 - f(X)] J_o(YX)$ and

$$\Omega = \Omega(t) = \begin{cases} [k_{\perp} w_c(t)]^2 & \text{for continuous surfaces,} \\ [\phi]^2 w_d(t)^2 & \text{for crystalline surfaces.} \end{cases}$$

The short-range and time-invariant structure factor during dynamic growth

For $\Omega(t) >> 1$, the diffraction structure factor $S(\mathbf{k}, t)$ is time-invariant and is determined by the short-range and time-independent parameters, α and η,

$$S(\mathbf{k}, t) \sim \begin{cases} (\eta k_{\perp}^{-\frac{1}{\alpha}})^2 F_{\alpha}(k_{\|} \eta k_{\perp}^{-\frac{1}{\alpha}}) & \text{for continuous surface} \\ (\eta [\phi]^{-\frac{1}{\alpha}})^2 F_{\alpha}(k_{\|} \eta [\phi]^{-\frac{1}{\alpha}}) & \text{for crystalline surfaces} \end{cases}$$

Note that for crystalline surfaces, the diffraction structure factor will deviate significantly from the above equation in the vicinity of out-of-phase conditions due to the discrete lattice effect. At near the out-of-phase diffraction condition, a more complicated form of S(**k**, t) can be obtained by combining Eq. (3.3) with the rigorous height difference function, Eq. (2.22). It can be shown that S(**k**, t) is still time-invariant for this exact result.

REFERENCES

5.1 H.-N. Yang, T.-M. Lu, and G.-C. Wang, *Phys. Rev. Lett.* **68**, 2612 (1992); *Phys. Rev.* **B47**, 3911 (1993).

5.2 Y.-L. He, H.-N. Yang, T.-M. Lu, and G.-C. Wang, *Phys. Rev. Lett.* **69**, 3770 (1992).

5.3 For $d = 1+1$, see, C. S. Lent and P. I. Cohen, *Surf. Sci.* **139**, 121 (1984); J. M. Pimbley and T.-M. Lu, *J. Vac. Sci. Technol.* **A2**, 457 (1984); J. M. Pimbley and T.-M. Lu, *J. Appl. Phys.* **57**, 1121 (1985).

5.4 For $d = 2+1$, see, H.-N. Yang, K. Fang, G.-C. Wang, and T.-M. Lu, *Europhys. Lett.* **19**, 215 (1992).

5.5 S. K. Sinha, E. B. Sirota, S. Garoff, and H. B. Stanley, *Phys. Rev.* **B38**, 2297 (1987).

5.6 Elliott A. Eklund, R. Bruinsma, J. Rudnick, and R. Stanley Williams, *Phys. Rev. Lett.* **67**, 1759 (1991).

5.7 J. Chevrier, V. Le Thanh, R. Buys, and J. Derrien, *Europhys. Lett.* **16**, 732 (1991).

5.8 G. L. M. K. S. Kahanda, Xiao-qun Zou, R. Farrell, and P.-Z. Wong, *Phys. Rev. Lett.* **68**, 3741 (1992).

5.9 Y.-N. Yang, Y.-S. Luo, and J. H. Weaver, *Bull. Am. Phys. Soc.* **37**, No.1, 666 (1992).

Chapter VI DIVERGENT INTERFACES

In contrast to the surface morphology discussed in the previous chapters where the interface width is finite, there is a very interesting case where one has a diverging interface. It occurs when the surface undergoes a roughening transition above a roughening temperature. This Chapter describes the salient features of the height-height correlation function and the structure factor of a diverging interface above the roughening temperature.

§VI.1 Correlation Functions during Surface Roughening Transition

In the roughening transition, a dramatic change in the correlation functions can be found.[6.1, 6.2, 6.3] For example, the non-divergent height-height correlation is transformed into a divergent one. Here we are dealing with an equilibrium surface structure.

§VI.1.1 *Surface roughening transition in crystalline surfaces*

Some of the defects that may occur at a free surface shown schematically in Fig. 6.1 are based on the terrace-ledge-kink (TLK) model.[6.4, 6.5] Among those defects are terrace ledges (or atomic steps), kinks, adatoms and vacancies. These defects can also occur in growth fronts discussed in previous chapters. The terraces are the exposed portions of low-Miller-index planes bonded by atomic steps. The kinks are the occasional jogs along a step ledge. The creation of kinks meanders a straight step ledge. The adatoms are the extra atoms that sit on the top layer of a surface. The vacancies refer to those lattice sites in a crystal surface that are not occupied by atoms.

The change of the surface morphology as a function of temperature can be shown schematically in Fig. 6.2. [6.1] At low temperatures, $T < T_R$, where T_R is the surface roughening transition temperature, the equilibrium surface contains a few thermally excited defects and exhibits an atomically flat morphology. At high temperatures, $T \geq T_R$, as shown in Fig. 6.2, with the proliferation of a large number of thermally excited steps, the low-temperature flat surface has been transformed into a rough and disordered structure where multilayers of atoms are involved to produce a mountain-valley-like landscape.

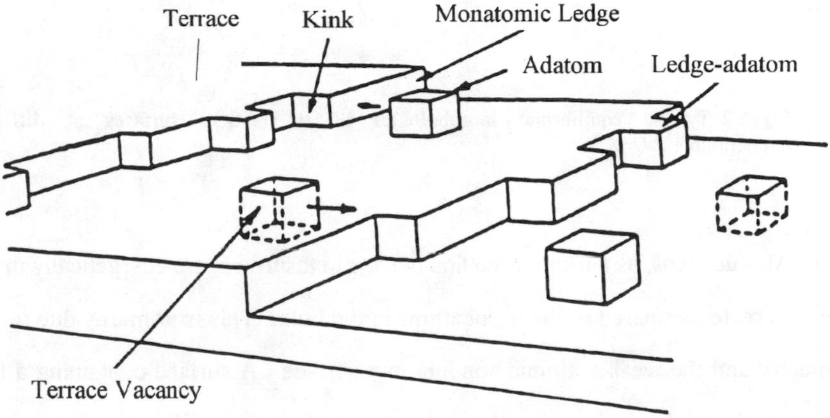

Fig. 6.1 A schematic drawing of defects that may occur at a free surface. [6.4, 6.5]

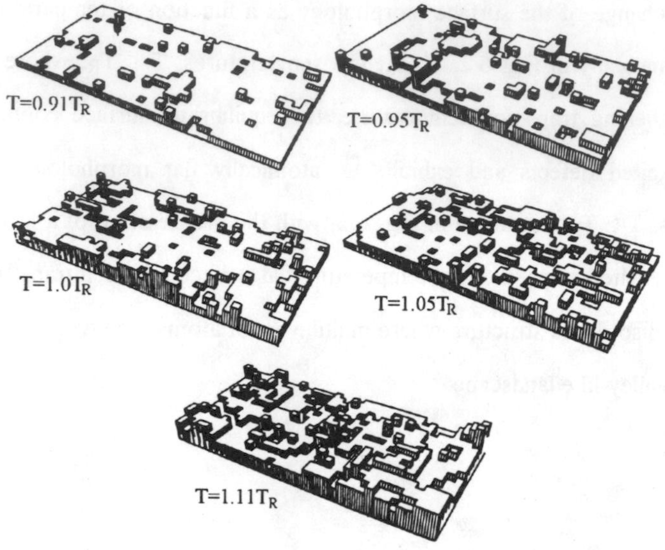

Fig. 6.2 Typical equilibrium morphologies of (001) SOS surfaces at different temperatures. [6.1]

Atomic steps, as a form of the line defects in a surface, are energetically much easier to create compared to the dislocations in the bulk. This is primarily due to the asymmetry and the weaker atomic bonding in a surface. A surface containing a few terraces does not change its ordered and flat morphology (it is still in the ordered-flat (OF) phase). However, if the number of terraces increases considerably, it will increase dramatically the ways to construct a roughed surface structure. The entropy gain can reduce significantly the surface free energy at high temperatures. The ordered flat phase then becomes thermally unstable. This instability leads to a catastrophic change of the surface perfection, i.e., the roughening transition in which the energy-stabilized OF phase is replaced by the entropy-stabilized rough phase.

The surface roughening transition usually refers to those that occur in the low-Miller-index surfaces. For a high-Miller-index surface (or vicinal surface), which intrinsically has equally spaced steps at low temperatures, the roughening manifests itself in the dramatic step meandering or the proliferation of kinks within those steps. [6.6, 6.7] (For distinction, we call it the "step roughening".)

The idea of the surface roughening was first proposed by Burton, Cabrera and Frank based on a two-dimensional (2D) lattice gas model. [6.8] Since then, many theories based on different step interactions have been developed. Among them, the discrete Gaussian (DG) model [6.9] proposed by Chui and Weeks is perhaps the most conclusive and generally accepted theory to describe the roughening transition.

§VI.1.2 *The height-height correlation function at the roughening transition*

Above the roughening transition temperature, T_R, the surface morphology shown in Fig. 6.2 has two important characteristics: (1) it contains a large number of terraces and steps; (2) it has a divergent solid-gas interface in which disordering can occur to an arbitrary depth in the crystal.

The general features of the surface roughening phase transition have been discussed in the review papers given by Weeks and Gilmer, [6.1] van Beijeren and Nolden, [6.2] and recently by Conrad. [6.3] The roughening transition is in the same universality class as that of the Kosterlitz-Thouless (KT) type [6.10] and is of infinite order. As T approaches T_R from below, the lateral correlation length and the interface width diverge very rapidly and remain infinite for $T \geq T_R$. A critical line extends from the roughening temperature T_R to the melting temperature T_m.

In a KT-roughened surface, the height-height correlation function diverges logarithmically [6.9] as $r \to +\infty$ at $T \geq T_R$,

$$H_d(\mathbf{r}) = \langle [h(\mathbf{r}) - h(0)]^2 \rangle \sim \frac{1}{\pi K_g} \mathrm{Ln}(r/a) \quad (r >> a,\ T \geq T_R), \qquad (6.1)$$

where a is the lattice spacing and the parameter K_g^{-1} depends on the temperature T. Since $H_d(r) \propto \mathrm{Ln}\, r$, the roughness exponent α is zero. The temperature dependent constant K_g is also a measure of the surface roughness. K_g has the universal value of $\pi/2$ at the KT transition temperature T_R and decreases monotonically as T increases above T_R.

In Appendix VIA, we give a full derivation of a 1D stepped surface that gives a rough surface based on the restricted solid-on-solid model. It is shown that this model is equivalent to the 1D Markovian surface proposed by Lu and Lagally [6.11] which has been used quite frequently in the literature for surface step analysis. These model surfaces contain unrestricted number of levels of steps and have a height-height correlation that diverges linearly with r. This model surface is therefore much rougher than the KT rough surface in 2D where the height-height correlation function only diverges logarithmically.

§VI.1.3 *Height difference function of the KT-rough phase*

At $T \geq T_R$, the logarithmically diverging height-height correlation, Eq. (6.1), leads to a power-law form of the height difference function. If the diffraction phase factor ϕ

is not too close to the "out-of-phase" condition, i.e., $|[\phi]| \neq \pi$, the height difference function can be derived by combining Eq. (6.1) with Eq. (2.23),

$$C_o(k_\perp, \mathbf{r}) \approx e^{-\frac{1}{2}[\phi]^2 H_d(\mathbf{r})} \approx r^{-\frac{1}{2\pi K_g}[\phi]^2} = r^{-\eta(\phi)}, \quad (r >> a, |[\phi]| \neq \pi). \qquad (6.2a)$$

where the exponent is given by

$$\eta(\phi) = \frac{1}{2\pi K_g}[\phi]^2. \qquad (6.3a)$$

However, Eq. (6.2a) does not apply near the out-of-phase condition due to the discrete lattice effect discussed in §II.2.2. We have shown that at $|[\phi]| \sim \pi$, the lowest order of $C_d(k_\perp, \mathbf{r})$ can be expressed by Eq. (2.23'),

$$C_o(k_\perp, \mathbf{r}) \approx e^{-\frac{1}{2}[\phi]^2 H_d(\mathbf{r})} + e^{-\frac{1}{2}(2\pi-|[\phi]|)^2 H_d(\mathbf{r})}, \quad (|[\phi]| \sim \pi),$$

under the condition of $H_d(\mathbf{r}) \geq 1$. Thus, referring to Eq. (6.1), we obtain the height difference function at the near out-of-phase condition as

$$C_o(k_\perp, \mathbf{r}) \approx r^{-\frac{1}{2\pi K_g}[\phi]^2} + r^{-\frac{1}{2\pi K_g}(2\pi-|[\phi]|)^2}$$

$$= r^{-\eta(\phi)} + r^{-\eta_1(\phi)}, \qquad (r >> a, |[\phi]| \sim \pi), \qquad (6.2b)$$

which is a sum of two power-law functions with different exponents, $\eta(\phi)$ and $\eta_1(\phi) = \frac{1}{2\pi K_g}(2\pi-[\phi])^2$. A particularly interesting case is at the exact out-of-phase condition, $\phi = k_\perp t = (2m-1)\pi$, $m = 0, \pm 1, \pm 2, \ldots$. At $\phi = (2m-1)\pi$, i.e., $|[\phi]| = \pi$, Eq. (6.2b) reduces to a single power-law function similar to Eq. (6.2a), except a factor of two difference,

$$C_0[\frac{(2m-1)\pi}{c}, \mathbf{r}] \approx 2\,r^{-\frac{\pi}{2K_g}} = 2\,r^{-\eta}, \qquad (r >> a).$$

The corresponding exponent is given by

$$\eta = \frac{\pi}{2K_g}, \qquad (6.3b)$$

which is equal to 1 at the transition temperature, $T = T_R$, since $K_g = \pi/2$.

§VI.2 Diffraction From A Surface Undergoing Roughening Transition

In the following, we give a full description of the diffraction characteristics from a surface undergoing thermal roughening transition.

§VI.2.1 *Diffraction structure factor from KT-rough phase*

According to Eq. (2.16), the Fourier transform of Eq. (6.2a) gives the diffraction structure factor from a KT-rough surface,

$$S(\mathbf{k}_{\|}, k_{\perp}) = \int d^2r\, C_0(k_{\perp}, \mathbf{r})\, e^{i\mathbf{k}_{\|}\cdot\mathbf{r}} = \int d^2r\, r^{-\eta(\phi)}\, e^{i\mathbf{k}_{\|}\cdot\mathbf{r}}$$

$$= \int_0^{\infty} r^{1-\eta(\phi)}\, dr \int_0^{2\pi} d\theta\, e^{ik_{\|}r\cos\theta}$$

$$= 2\pi \int_0^{\infty} r^{1-\eta(\phi)}\, J_0(k_{\|}r)\, dr, \qquad (|[\phi]| \neq \pi).$$

In the last step, if we replace the variable r in the integral by $x/k_{\|}$, we obtain,

$$S(\mathbf{k}_{\|}, k_{\perp}) = (k_{\|})^{-2+\eta(\phi)}\, 2\pi \int_0^{\infty} x^{1-\eta(\phi)}\, J_0(x)\, dx \propto (k_{\|})^{-2+\eta(\phi)}, \quad (|[\phi]| \neq \pi),$$

where the integral $\int_0^\infty x^{1-\eta(\phi)} J_0(x)dx$ is independent of $k_{||}$. A similar calculation can be applied to Eq. (6.2b) for the near out-of-phase condition. A general expression for the diffraction structure from a KT-roughened surface can then be written as

$$S(\mathbf{k}_{||}, k_\perp) \propto \begin{cases} (k_{||})^{-2+\eta(\phi)} & \text{away from the out-of-phase condition ,} \\ 2(k_{||})^{-2+\eta} & \text{at the out-of-phase condition, } ||[\phi]|| = \pi \text{ ,} \end{cases} \qquad (6.4)$$

where $\eta(\phi)$ and η are given by Eqs. (6.3a) and (6.3b), respectively.

The power-law form of the diffraction structure factor from a KT-roughened surface, as shown in Eq. (6.4), is distinctly different from the structure factors for a rough surface with a non-divergent interface width discussed in Chapter III and Chapter V. From the out-of-phase to the near in-phase diffraction condition, the line shape of Eq. (6.4) has only one component, the power-law function. In contrast, for a non-divergent surface, the line shape contains both a δ–function and a diffuse component. The non-existence of the δ–component is typical of a rough surface with a divergent interface width. We also note that for a divergent 1D Markovian surface (see Appendix VIA), the diffraction structure factor contains only one component, the 1D Lorentzian function given by Eq. (ApVIA.12).

The power-law line shape shown in Eq. (6.4) has an exponent, $2 - \eta(\phi)$. At the out-of-phase condition, the exponent $= 2 - \frac{\pi}{2K_g}$, and is equal to 1 at the roughening transition temperature. The exponent decreases monotonically for $T > T_R$.

Experimentally, one can measure the tail part of the diffraction intensity distribution as a function of $\mathbf{k}_{\|}$. The exponent $\eta(\phi)$ can then be extracted from the plot of $\mathrm{Ln}[S(\mathbf{k}_{\|}, k_{\perp})]$ vs. $\mathrm{Ln}(k_{\|})$, where the slope of the plot is equal to $-2 + \eta(\phi)$.

§VI.2.2 *Consideration of short-distance roughness*

We should be aware of the fact that the divergent height-height correlation function, shown in Eq. (6.1), only describes the asymptotic behavior (large distance) of the KT-roughened surface. Thus, the predicted power-law line shape for the diffraction intensity, Eq. (6.4), only reflects the global feature of the rough surface. Equation (6.4) alone may not be sufficient to interpret all the experimental data. This is because short-distance roughness, which reflects the local defect structure, always exists and makes a certain contribution to the diffraction profile. Short-distance roughness for the divergent rough surface contributes a broad diffuse background to the diffraction line shape. In a manner analogous to the "δ + diffuse" line shape in a non-divergent surface, the line shape for a KT-roughened surface can be described by the "power-law + diffuse" profile. Since most theoretical models do not provide explicit solutions to the short-distance correlation functions, establishing a phenomenological model becomes an alternative approach. Experimental observations [6.12 - 6.16] suggested that the diffuse profile can usually be described by a Lorentzian line shape. Such a conclusion is also consistent with simulation results. [6.17] Therefore, phenomenologically, the diffraction intensity distribution from a KT-roughened surface can be described by a line shape of "power-law + Lorentzian".

§VI.3 Roughening Transitions in Metal Surfaces

Thermal roughening phase transitions have been observed in both high-Miller-index and low-Miller-index metal surfaces. The high-Miller-index surfaces (or vicinal surfaces) intrinsically have equally spaced terraces with atomic steps at T = 0 K and are capable of undergoing a step roughening transition below the melting temperature, T_m. Examples are (11*m*) surfaces of Ni [6.7, 6.12, 6.18, 6.19] and Cu. [6.6, 6.14, 6.20 - 6.22] The low-Miller-index metal surfaces have no intrinsic steps and are more stable. Among them, the surfaces of FCC(110) face are less stable than those of FCC(100) and FCC(111) faces. It is expected that a possible thermal roughening transition below T_m could occur in some of FCC(110) metal surfaces. [6.23] Examples have been found in Ag(110), [6.13, 6.24] Pb(110), [6.15, 6.25] Ni(110) [6.16] and Pt(110) [6.26] surfaces. Cu(110) is also a possible candidate [6.27] although some disputes still exist. [6.28] For low-Miller-index metal surfaces other than the FCC(110) face, only one surface, Pt(100), has been reported to become thermally rough below T_m. [6.29]

The major experimental approach for the study of thermal roughening phenomenon is to use diffraction techniques, such as HRLEED, X-ray grazing angle scattering, and atomic scattering. Determination of thermal roughening in a surface comes from a line shape analysis based on the characteristics of the power-law type of diffraction structure factor shown in §VI.2.1. The power-law line shape has been found in many high-Miller-index and low-Miller-index metal surfaces. These are clear evidences for the existence of logarithmically diverging height-height correlation. However, in some diffraction experiments, line shapes other-than power-law types have also been observed in high-temperature rough phases. [6.12, 6.16, 6.24, 6.26] This fact suggests that there may exist rough surfaces other than KT type.

(Recall that a KT roughened surface has a logarithmically divergent height-height correlation function and gives a power-law type of line shape.) Another possibility is that the measured line shapes include other effects, such as contributions from the one-phonon scattering [6.3] or the short-distance roughness discussed in §VI.2.2. Under these circumstances, the analysis of diffraction line shape is not adequate. There can be multiple interpretations of experimental data.

§VI.4 Summary of Correlation Functions and Diffraction Line Shapes from Different Rough Surfaces

We have shown in this chapter that rough surfaces with divergent interface widths have distinctly different height-height correlation functions, height difference functions, and diffraction structure factors as compared to those of non-divergent interfaces. For comparison, we show schematically these differences in Table VI-1. The results for perfectly flat surfaces have been discussed in Chapter I. The features for non-divergent rough surfaces are refereed to Chapters II and III. The results of KT rough surfaces with a divergent interface width are discussed in previous sections of this Chapter. For Markovian surfaces, interested readers are refereed to Appendix VIA.

Table VI-1. Comparison of Correlation Functions and Diffraction Line Shapes for Different Types of Surfaces Considered in This Monograph

Surface Morphology	**Height-height Correlation** $< [z(\mathbf{r}) - z(0)]^2 >$	**Height Difference Function** $C(k_\perp, \mathbf{r})$: $< e^{ik_\perp[z(\mathbf{r}) - z(0)]} >$	**Diffraction Line Shape** $\int d\mathbf{r}\, C(k_\perp,\mathbf{r})\, e^{ik_\parallel \cdot \mathbf{r}}$
Perfectly flat surface (Chapter I)	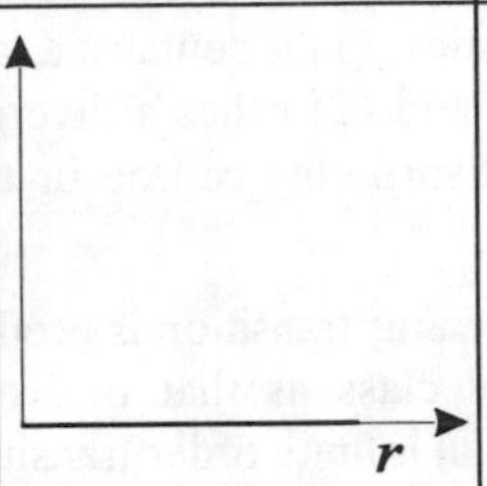	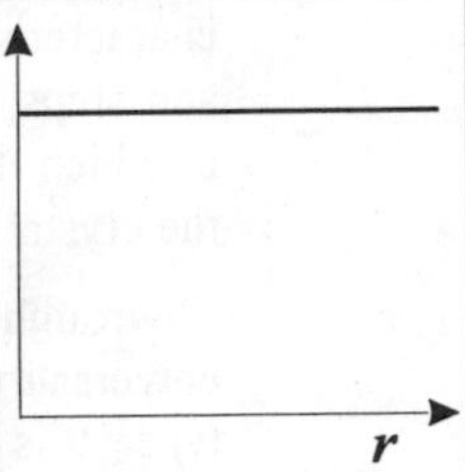	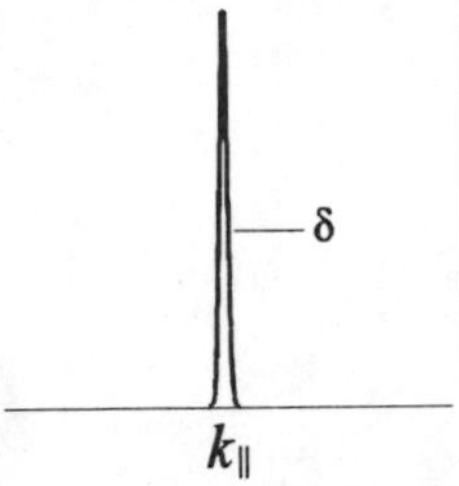
Rough surface with finite interface width (Chapters II, III)	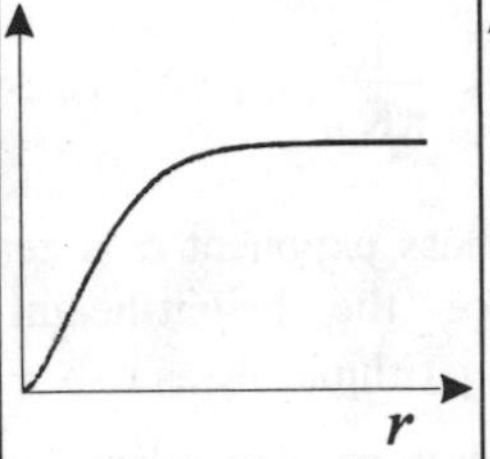	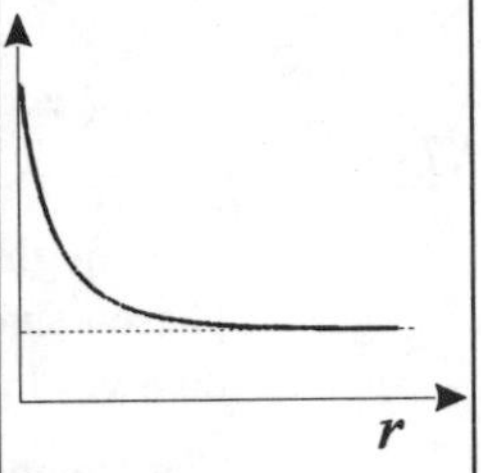	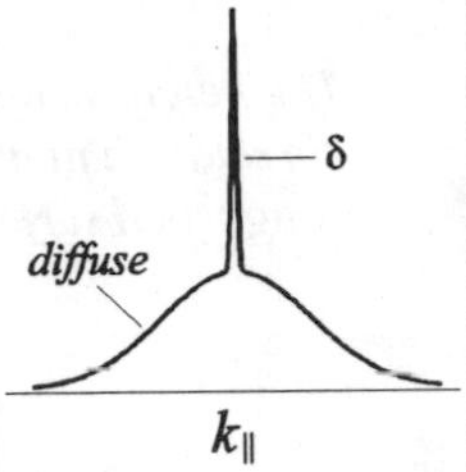
KT rough surfaces with logarithmically divergent height-height correlation (Chapter VI)	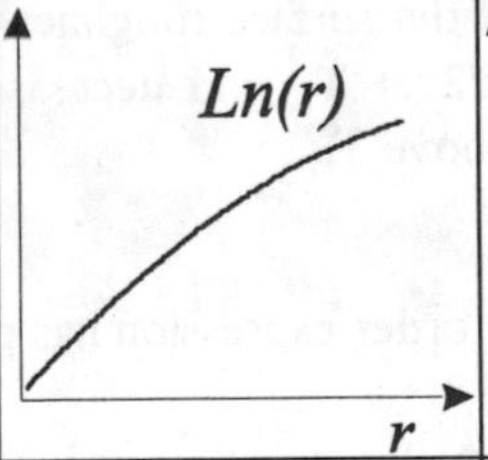	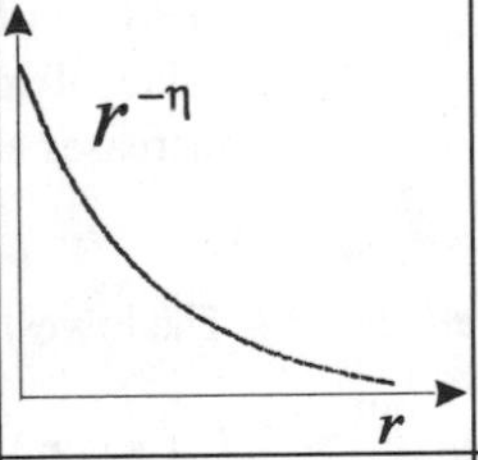	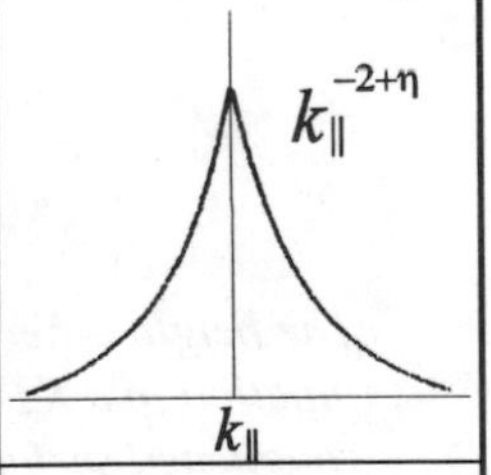
Markovian surfaces with linearly divergent height-height correlation (Appendix VIA)	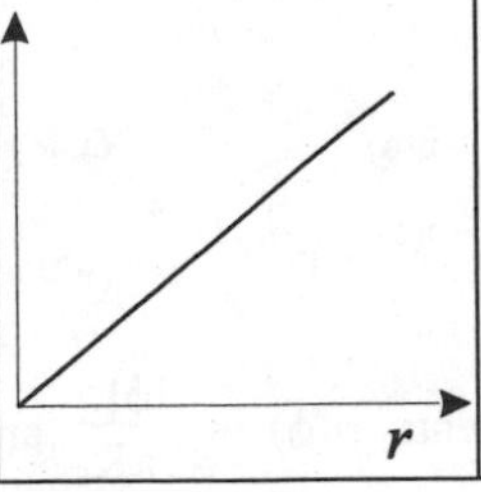	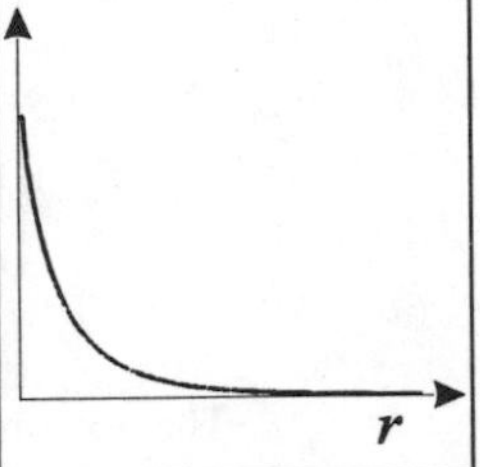	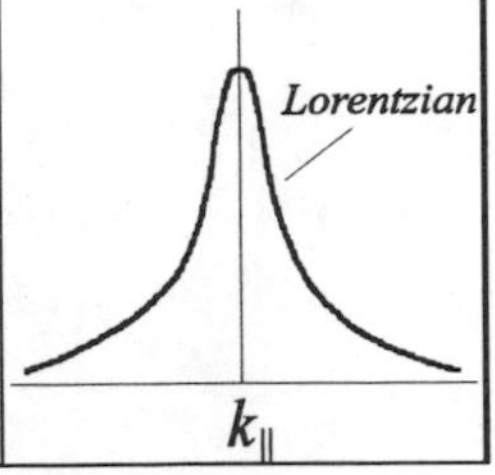

REVIEW AND SUMMARY

Equilibrium surface roughening transition

Thermal roughening can occur in a surface at certain temperature, T_R, at which the flat surface morphology becomes thermally unstable and this instability leads to a catastrophic change in surface perfection.

A thermally roughened surface has two main characteristics: (1) it contains a large number of terraces and steps; and (2) it has a divergent solid-gas interface in which disordering can occur to an arbitrary depth in the crystal.

The roughening transition is predicted to be in the same universality class as that of Kosterlitz-Thouless (KT) type. It is an infinite order transition.

The height-height correlation in a KT-roughened surface

$$H_d(\mathbf{r}) \sim \frac{1}{\pi K_g} \mathrm{Ln}(r/a) \quad (r >> a,\ T \geq T_R). \qquad (6.1)$$

The roughness exponent α is zero for a KT-roughened phase since the height-height correlation function diverges logarithmically as $r \to +\infty$.

The temperature dependent constant K_g is also a measure of the surface roughness. K_g has the universal value of $\pi/2$ at T_R and decreases monotonically as T increases above T_R.

The height difference function in a KT-roughened surface

The lowest order expression has power-law forms:

$C_d(k_\perp, \mathbf{r}) \sim$

$$\begin{cases} r^{-\eta(\phi)} & (r >> a,\ \|[\phi]\| \neq \pi), \\ r^{-\eta(\phi)} + r^{-\eta_1(\phi)} & (r >> a,\ \|[\phi]\| \sim \pi). \end{cases} \qquad (6.3)$$

The exponents $\eta(\phi) = \dfrac{[\phi]^2}{2\pi K_g}$ and $\eta_1(\phi) = \dfrac{1}{2\pi K_g}(2\pi - \|[\phi]\|)^2$.

Diffraction structure factors in a KT-roughened surface

A power-law type of the diffraction structure factor can be derived from Eq. (6.3),

$$S(\mathbf{k}_{\|}, k_{\perp}) \propto \begin{cases} (k_{\|})^{-2+\eta(\phi)} & (|[\phi]| \neq \pi), \\ 2(k_{\|})^{-2+\eta} & (|[\phi]| = \pi), \end{cases} \tag{6.4}$$

where $\eta = \dfrac{\pi}{2K_g}$, which is equal to 1 at $T = T_R$ and increases monotonically at $T > T_R$.

The diffraction structure factor from a KT-roughened surface has only the power-law type of diffuse profile, which is distinctly different from the structure factor for a non-divergent rough surface, where the line shape contains both a δ–function and a diffuse component. The non-existence of the δ–component is typical of a rough surface with a divergent interface width.

Appendix VIA One-dimensional Markovian Surface

The one-dimensional Markovian surface is one of the few models of the multi-level crystalline surface which has an analytical solution for the diffraction problems. [6.11] In this appendix, we discuss the important properties of this model surface.

1. *One-dimensional Markovian surface*

Figure ApVIA.1 is the schematic of a one-dimensional stepped surface, where steps occur only in one particular direction on a two-dimensional surface. For simplicity, we assume the steps have only a monatomic height, i.e., a single step height, c. For the purpose of defining a probability of meeting a step we assume for now the simplest lattice, namely one in which atoms in different layers sit directly over each other, .i.e., AAA stacking. The displacement at any step is then given by $\delta z = \pm c$. The displacement may be either "upward" (+) or "downward" (−), with equal probability.

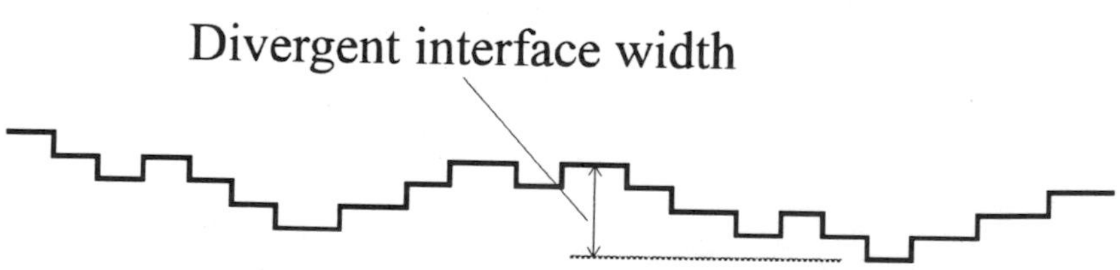

Fig. ApVIA.1 Schematic of a one-dimensional divergent stepped surface.

Let γ be the probability of meeting an atom displaced vertically either upward or downward in going from any lattice site to an adjacent one. γ is thus the probability to meet a step and $1 - \gamma$ is that of meeting an atom in the terrace (not at steps). Since each displacement or step occurs independent of the other, this one-dimensional step surface shown in Fig. ApVIA.1 is similar to the path of a random walk, or a one-dimensional Markovian chain. For the present random walk model, the displacement at each walk has three choices, an upward walk of a displacement $\boldsymbol{a} + \boldsymbol{c}$, with a probability, $\gamma/2$; a downward walk of a displacement $\boldsymbol{a} - \boldsymbol{c}$, with a probability, $\gamma/2$; and a lateral walk of a displacement $\boldsymbol{a}$, with a probability, $1 - \gamma$. The terrace-width distribution is then given by the geometric distribution,

$$\mathrm{P(M)} = (1 - \gamma)^{\mathrm{M}-1}\gamma, \qquad \text{(ApVIA.1)}$$

with an average terrace width,

$$< \mathrm{L} > = \sum_{\mathrm{M}=1}^{\infty} (\mathrm{M}a)\, \mathrm{P(M)} = \sum_{\mathrm{M}=1}^{\infty} (\mathrm{M}a)\, (1 - \gamma)^{\mathrm{M}-1}\gamma = \frac{a}{\gamma}. \qquad \text{(ApVIA.2)}$$

The present Markovian surface can be derived from a statistical mechanics theory which is shown next.

2. *One-dimensional restricted solid-on-solid model*

Consider a nearest-neighbor interaction Hamiltontian in a 1D lattice with a lattice spacing of a,

$$\mathrm{E} = \sum_{n} \frac{1}{2} \varepsilon\, |\mathrm{h}(x_n) - \mathrm{h}(x_n \pm a)|\, , \qquad \text{(ApVIA.3)}$$

where $x_n = na$, $n = 0, \pm 1, \pm 2, \ldots$, is the lateral site of the 1D lattice and $h(x_n)$ represents the surface height at x_n in unit of c, $z(x_n) = h(x_n)c$. In the solid-on-solid model, the surface height $h(x_n)$ is assumed to be a single valued function so that the possibility of the overhangs in the surface is excluded. Also, the value of the nearest-neighbor height difference, $h(x_n)-h(x_n \pm a)$, is restricted to be 0 and $\pm$ 1. Such restriction allows only single steps to exist in the surface. This restriction defines what is called the restricted solid-on-solid model. The nearest-neighbor coupling constant, $\varepsilon > 0$, represents the single step energy. As the result, Eq. (ApVIA.3) represents the total surface step energy in the 1D restricted solid-on-solid model. The surface has a ground state, $E = 0$, corresponding to a perfectly flat and step-free surface. In contrast, the surface containing steps has a higher surface energy, $E > 0$.

The thermal equilibrium behavior of the 1D restricted solid-on-solid model surface is governed by the partition function given by

$$Z = \sum_{\{h(x_n)\}} e^{-\beta E} = \sum_{\{h(x_n)\}} \exp\left(-\sum_n \frac{1}{2}\beta\varepsilon\,|h(x_n) - h(x_n \pm a)|\right), \qquad (\text{ApVIA.4})$$

where $\beta = \dfrac{1}{k_B T}$ and $\{h(x_n)\}$ denotes that the summation is over all possible values of the surface height, $h(x_n)$, $n = 0, \pm 1, \pm 2, \ldots$. Under our restriction, $h(x_n)-h(x_n+a) = 0$ or $\pm$ 1. If we define a new set of variables, $\Delta h_n = h(x_n)-h(x_n+a)$, $n = 0, \pm 1, \pm 2, \ldots$, the calculation of Eq. (ApVIA.4) can be easily carried out,

$$Z = \prod_n \left(\sum_{\Delta h_n = -1}^{1} e^{-\beta\varepsilon|\Delta h_n|}\right) = (1 + 2e^{-\beta\varepsilon})^N, \qquad (\text{ApVIA.4}')$$

where N is the total number of lattice sites.

The surface free energy per unit site is then given by

$$F = -\frac{\beta}{N} \text{Ln } Z = -\beta \text{ Ln}(1 + 2e^{-\beta\varepsilon}) . \qquad (\text{ApVIA.5})$$

The free energy, shown in Eq. (ApVIA.5), is an analytical function of the temperature except at T = 0 K. At T > 0, the step free energy F < 0. We will show later that the surface at T > 0 is in the rough phase with a divergent interface width. At T = 0 K, we have F = 0. T = 0 K is considered as the phase transition temperature.

It can be shown that the one-dimensional Markovian step surface discussed in the last section is equivalent to the one-dimensional restricted solid-on-solid model. According to Eqs. (ApVIA.3) and (ApVIA.4), the probability of meeting a step at a site of x_n is equivalent to the probability of the vertical displacement $\Delta h_n = 1$ or -1. The probability is therefore equal to $C^{-1}e^{-\beta\varepsilon|\Delta h_n|} = C^{-1}e^{-\beta\varepsilon}$, where C^{-1} is a normalized prefactor. On the other hand, the probability of meeting a site of x_n in the terrace (not at a step) is equal to the probability for no displacement in the *z*-direction, $\Delta h_n = 0$. The corresponding probability is then given by $C^{-1}e^{-\beta\varepsilon|\Delta h_n|} = C^{-1}$ The summation for the probabilities shown above should be equal to one, which then determines the normalized prefactor,

$$C = \sum_{\Delta h_n = -1}^{1} e^{-\beta\varepsilon|\Delta h_n|} = 1 + 2e^{-\beta\varepsilon} .$$

Recall that in the one-dimensional Markovian surface model, the parameter γ is defined as the probability to meet a step, i.e., the probability for $\Delta h_n = \pm 1$. γ therefore has an explicit form in the one-dimensional restricted solid-on-solid model,

$$\gamma = \frac{2e^{-\beta\varepsilon}}{1 + 2e^{-\beta\varepsilon}} . \qquad (\text{ApVIA.6})$$

The Hamiltonian in the 1D restricted solid-on-solid model, shown in Eq. (ApVIA.3), has only nearest-neighbor interaction terms. Therefore, whether or not the step can exist at the site of x_n is only determined by the probability, $C^{-1}e^{-\beta\varepsilon|\Delta h_n|}$, and does not depend on the steps at other sites. This important characteristic is consistent with the assumption of independent probabilities in the one-dimensional Markovian chain model.

3. *Correlation functions*

The calculation of the height difference function is relatively simple. In the expression, $C_d(k_\perp, \mathbf{r} = na) = \langle e^{ik_\perp c\,[h(na) - h(0)]} \rangle$, the height difference, $[h(na) - h(0)]$ can be represented as the sum of the height differences between successive sites from $\mathbf{r} = 0$ to $\mathbf{r} = na$,

$$h(na) - h(0) = [h(a) - h(0)] + [h(2a) - h(a)] +$$

$$+ [h(3a) - h(2a)] + \ldots + [h(na) - h((n-1)a)].$$

Because the probability γ of having a step at a site is assumed independent of that at other sites, the average of the product should be equal to the product of the average,

$$C_d(k_\perp, na) = \langle e^{ik_\perp c\,[h(na) - h(0)]} \rangle$$

$$= \langle e^{ik_\perp c\,[h(a) - h(0)]} \rangle \times \langle e^{ik_\perp c\,[h(2a) - h(a)]} \rangle \times$$

$$\times \langle e^{ik_\perp c\,[h(3a) - h(2a)]} \rangle \times \ldots \times \langle e^{ik_\perp t\,[h(na) - h((n-1)a)]} \rangle$$

$$= f^{|n|}, \qquad (\text{ApVIA.7})$$

where

$$f = \langle e^{ik_\perp c\,[h(a) - h(0)]} \rangle = \frac{\gamma}{2} e^{ik_\perp c} + \frac{\gamma}{2} e^{-ik_\perp c} + (1-\gamma)\, e^{i0} = 1 - \gamma\,[\,1 - \cos(k_\perp c)\,].$$

The height difference function in the one-dimensional Markovian surface can then be expressed as a decreasing exponential,

$$C_d(k_\perp, \mathbf{r}) = \{1 - \gamma\,[\,1 - \cos(k_\perp c)\,]\}^{|n|} = e^{-r\sigma_c} = e^{-\frac{r}{\eta_o}[1-\cos(k_\perp c)]}, \qquad (\text{ApVIA.7}')$$

where r = $|n|a$ and,

$$\sigma_c = -\frac{\mathrm{Ln}(f)}{a} = \frac{1-\cos(k_\perp c)}{\eta_o} = -\frac{\mathrm{Ln}\{1 - \gamma\,[\,1-\cos(k_\perp c)\,]\}}{a}. \qquad (\text{ApVIA.8})$$

Usually, $\gamma \ll 1$, i.e., the average terrace size $\langle L \rangle$ is much larger than the lattice spacing, a. For $\gamma \ll 1$, the above equation can be expressed approximately as

$$\sigma_c \approx \frac{\gamma}{a}[\,1 - \cos(k_\perp c)\,], \qquad (\text{ApVIA.8}')$$

i.e., $\eta_o \approx a/\gamma = \langle L \rangle$.

One may notice that the calculation of the height difference function, Eq. (ApVIA.7), does not employ the relative height distribution function, $g(\Delta z, \mathbf{r})$, due to the high symmetry of the one-dimensional Markovian chain model. The derivation of $g(\Delta z, \mathbf{r})$ can be accomplished using the generation function, $G_n(x) = [(1-\gamma) + \frac{\gamma}{2}x + \frac{\gamma}{2}x^{-1}]^n$. The distribution function, $g(\Delta z = mc, \mathbf{r} = na) = g(m, n)$, is the coefficient of the term x^m in the polynomial expression of $G_n(x)$, $G_n(x) = \sum_{m=-n}^{n} g(m, n)\, x^m$. If $|m| \le n$,

$$g(m, n) = \sum_{k=0}^{[(n-|m|)/2]} \frac{n!}{k!\,(k+|m|)!\,(n-|m|-2k)!} \left(\frac{\gamma}{2}\right)^{2k+|m|} (1-\gamma)^{n-|m|-2k}, \qquad \text{(ApVIA.9)}$$

while for $|m| > n$, $g(m, n) = 0$, where [ ] is defined as [N/2] = M for either N = 2M or N = 2M + 1. Using Eq. (ApVIA.9), one can easily prove that

$$C_d(k_\perp, n\mathbf{a}) = \langle e^{ik_\perp c\,[h(n\mathbf{a}) - h(0)]} \rangle = \sum_{m=-n}^{n} g(m, n)\, e^{ik_\perp c\, m} = G_n(\, x = e^{ik_\perp c}\,) = f^{|n|},$$

which is identical to the previous result, Eq. (ApVIA.7).

In the one-dimensional Markovian surface model, the relative height distribution function, Eq. (ApVIA.9), is rigorous but complicated. We can show that under certain conditions, Eq. (ApVIA.9) can be represented approximately by a Gaussian function.

According to the Fourier transform relation, Eq. (2.17'), we have

$$g(m, n) = \frac{1}{2\pi} \int_{-\pi}^{+\pi} d\phi\; C_d(k_\perp, \mathbf{r})\, e^{-i\phi\, m}$$

$$= \frac{1}{2\pi} \int_{-\pi}^{+\pi} d\phi\; e^{-\frac{r}{\eta_o}[1-\cos(\phi)]}\, e^{-i\phi\, m}, \qquad \text{(ApVIA.10)}$$

where $r = |n|a$. Using Eq. (ApVIA.10), one can simplify the relative height distribution function, $g(m, n)$, in the case of $r \gg \eta_o$. Since $r/\eta_o = n\gamma \gg 1$, the exponential function, $e^{-\frac{r}{\eta_o}[1-\cos(\phi)]}$, has non-zero values only at the vicinity of $\phi = 0$, where $e^{-\frac{r}{\eta_o}[1-\cos(\phi)]} \sim 1$. We can then expand $\cos(\phi) \approx 1 - \frac{1}{2}\phi^2$ and obtain,

$e^{-\frac{r}{\eta_o}[1-\cos(\phi)]} \approx e^{-\frac{r}{2\eta_o}\phi^2}$. Equation (ApVIA.10) thus becomes a Gaussian distribution function,

$$g(m, n) \approx \frac{1}{2\pi}\int_{-\infty}^{+\infty} d\phi\, e^{-\frac{r}{2\eta}\phi^2} e^{-i\phi m} = \frac{1}{\sigma\sqrt{2\pi}} e^{-\frac{m^2}{2\sigma^2}}, \quad (n\gamma >> 1), \qquad (\text{ApVIA.11})$$

where $\sigma^2 = r/\eta_o = n\gamma$ is the asymptotic height-height correlation function in this one-dimensional Markovian surface.

One may notice that the quantity, $n\gamma = \sigma^2$, is the average number of steps within a region of *na*. Equation (ApVIA.11) therefore implies that the Gaussian distribution function can be a good approximation in a region which is sufficiently rough so that $n\gamma >> 1$.

It is easy to calculate the height-height correlation function from Eq. (ApVI.A.7) as

$$H(\mathbf{r}) = -\frac{d^2}{d\phi^2}\bigg|_{\phi=0} C_d(k_\perp, \mathbf{r}) = \frac{r}{\eta_o} = n\gamma = \sigma^2, \qquad (\text{ApVIA.12})$$

which is consistent with the asymptotic result ($r/\eta_o >> 1$) obtained from Eq. (ApVIA.11).

Equation (ApVIA.12) has a linear relation between H(r) and the distance r, which gives a roughness parameter, $\alpha = 0.5$. A similar characteristic also appears for a general random walk or a Brownian motion problem. This property is related to a class of self-affine fractals which exist in natural and geometrical objects.

H(**r**) becomes arbitrarily large if $r \to \infty$. The amplitude of the surface height fluctuation thus diverges for a very large system. Both the one-dimensional Markovian surface and the one-dimensional restricted solid-on-solid surface belong to the category of rough surfaces that have divergent interface width as defined by the divergent height-height correlation function, $H(\mathbf{r}) \to \infty$ for $r \to \infty$.

The surface height fluctuation also depends on the step density, γ, because the linear relation in Eq. (ApVIA.12) has a slope, $1/\eta_0$, which is proportional to γ. If the surface is very flat, i.e., the step density $\gamma \to 0$, the surface height fluctuation will then vanish because $H(\mathbf{r}) \to 0$ as $\gamma \to 0$. On the other hand, if $\gamma \to 0$, we have $T \to 0$ according to Eq. (ApVIA6). This implies that in the one-dimensional restricted solid-on-solid surface, the ordered-flat phase can only exist at T = 0 K. T = 0 K is therefore considered as the roughening transition temperature.

4. *Diffraction structure factor*

It is easy to calculate the diffraction structure factor for the one-dimensional Markovian surface. Combining Eq. (ApVIA.7) with Eq. (2.11), we can obtain the discrete version of the diffraction structure factor. Also, if we combine Eq. (ApVIA.7') with Eq. (2.16), the continuous version of the diffraction structure factor can be obtained. The results can be summarized in the following equations,

$$S(\mathbf{k}_{\|}, k_{\perp}) \sim \begin{cases} \dfrac{(1-f)^2}{1+f^2-2f\cos(k_{\|}a)} & \text{(discrete version)}, \\ \dfrac{1}{1+|k_{\|}|^2/\sigma_c^{\,2}} & \text{(continuous version)}, \end{cases} \qquad \text{(ApVIA.13)}$$

where σ_c is given by Eq. (ApVIA.8) and f is given by $f = 1 - \gamma[1-\cos(k_{\perp}c)]$. The corresponding FWHM measured in $k_{\|}$ can be calculated as

$$\text{FWHM} = \begin{cases} \dfrac{2}{a}\arccos\left(1-\dfrac{(1-f)^2}{2f}\right) & \text{(discrete version)}, \\ 2\sigma_c & \text{(continuous version)}. \end{cases}$$

We note that the diffraction line shape in the continuous version is given by a 1D Lorentzian function. In the case of $\gamma << 1$, i.e., $f \sim 1$ and $0 < 1 - f << f$, the FWHM for the discrete version becomes

$$(\mathrm{FWHM})_{\mathrm{discrete}} \approx 2\,\frac{1-f}{a\sqrt{f}} \approx 2\,\frac{\gamma}{a}\,[\,1-\cos(k_{\perp}c)\,]$$

$$\approx 2\sigma_c = (\mathrm{FWHM})_{\mathrm{continuous}}\,, \qquad (\text{ApVIA.14})$$

where the approximation of σ_c at $\gamma << 1$ is given by Eq. (ApVIA.8'). We thus conclude that the FWHM is about the same at $\gamma << 1$ for both discrete and continuous versions. The physical reason for this is that the average terrace size at $\gamma << 1$ is large enough so that the discrete atomic nature of the terrace can be ignored in the diffraction experiment.

The relation of FWHM vs. $k_{\perp}c$, as given by Eq. (ApVIA.14), shows an oscillatory behavior with a period of 2π in $\phi = k_{\perp}c$. The FWHM takes a maximum value at the out-of-phase condition, $\phi = k_{\perp}c = (2m-1)\pi$, and a minimum value at the in-phase condition, $\phi = k_{\perp}c = 2m\pi$. The oscillatory period can thus be used experimentally to determine the atomic step height, c. On the other hand, from Eq. (ApVIA.14), we see that the amplitude of the oscillatory FWHM at the out-of-phase condition is given by $4\frac{\gamma}{a}$, which is proportional to the step density γ (or inversely proportional to the average terrace size, $\langle L \rangle = a/\gamma$). Experimentally, γ can be determined from the FWHM of the line shape measured at the out-of-phase condition.

REFERENCES

6.1 For reviews see, J. D. Weeks and G. H. Gilmer, in *Advances in Chemical Physics*, Vol. **XL**, (John Wiley & Sons, New York, 1979), pp. 157; J. D. Weeks, in *Ordering of Strongly Fluctuating Condensed Matter System* (Plenum, New York, 1980), pp. 293.

6.2 Also see, H. van Beijeren and I. Nolden, in *Structures and Dynamics of Surface* II, *Topics in Current Physics* Vol. **43**, (Springer - Verlag, Berlin, 1987), pp. 259.

6.3 For a recent review see, E. H. Conrad, *Progress in Surf. Sci.* **39**, 65 (1992).

6.4 W. Kossel, Nachr. Ges. Wiss. Goetting, *Math.-Phys.* **Kl** (1927), p.185.

6.5 I. N. Stranski, *Z. Phys. Chem.* **136**, 259 (1928).

6.6 J. Villain, D. R. Grempel, and J. Lapujoulade, *J. Phys.* **F15**, 809 (1985).

6.7 M. den Nijs, E. K. Riedel, E. H. Conrad, and T. Engel, *Phys. Rev. Lett.* **55**, 1689 (1985).

6.8 W. K. Burton, N. Cabrera, and F. C. Frank, *Philos. Trans. Roy. Soc. London* **243A**, 299 (1951).

6.9 S. T. Chui and J. D. Weeks, *Phys. Rev.* **B14**, 4978 (1976).

6.10 J. M. Kosterlitz and D. J. Thouless, *J. Phys.* **C6**, 1181 (1973); J. M. Kosterlitz, ibid, **C**7, 1046 (1974).

6.11 T. -M. Lu and M. G. Lagally, *Surf. Sci.* **120**, 47 (1982).

6.12 I. K. Robinson, E. H. Conrad, and D. S. Reed, *J. Phys.* (*Paris*) **51**, 103 (1990).

6.13 G. A. Held, J. L. Jordan-Sweet, P. M. Horn, A. Mak, and R. J. Birgeneau, *Phys. Rev. Lett.* **59**, 2075 (1987).

6.14 B. Salanon, F. Fabre, J. Lapujoulade, and W. Selke, *Phys. Rev.* **B38**, 7385 (1988).

6.15 H.-N. Yang, T.-M. Lu, and G.-C. Wang, *Phys. Rev. Lett.* **63**, 1621 (1989); *Phy. Rev.* **B43**, 4714 (1991).

6.16 Y. Cao and E. H. Conrad, *Phys. Rev. Lett.* **64**, 447 (1990).

6.17 N. C. Bartelt, T. L. Einstein and Ellen Williams, *Surf. Sci.* **276**, 308 (1992).

6.18 E. H. Conrad, L. R. Allen, D. L. Blanchard, and T. Engel, *Surf. Sci.* **187**, 265 (1987).

6.19 D. L. Blanchard, D. F. Thomas, H. Xu, and T. Engel, *Surf. Sci.* **222**, 477 (1989).

6.20 K. S. Liang, E. B. Sirota, K. L. D'Amico, G. J. Hughess, and S. K. Sinha, *Phys. Rev. Lett.* **59**, 2447 (1987).

6.21 F. Fabre, D. Gorse, B. Salanon, and J. Lapujoulade, *J. Physique* (*Paris*) **48**, 1017 (1987).

6.22 F. Fabre, D. Gorse, and J. Lapujoulade, and B. Salanon, *Europhys. Lett.* **3**, 737 (1987).

6.23 A. Trayanov, A. C. Levi, and E. Tosatti, *Europhys. Lett.* **8**, 657 (1989); *Surf. Sci.* **233**, 184 (1990).

6.24 I. K. Robinson, E. Vlieg, H. Hornis, and E. H. Conrad, *Phys. Rev. Lett.* **67**, 1890 (1991).

6.25 J. C. Heyraud and J. J. Metois, *J. Crys. Growth* **82**, 269 (1987).

6.26 I. K. Robinson, E. Vlieg, and K. Kern, *Phys. Rev. Lett.* **63**, 2578 (1989).

6.27 S. G. J. Mochrie, *Phys. Rev. Lett.* **59**, 304 (1987).

6.28 P. Zeppenfeld, K. Kern, R. David, and G. Comsa, *Phys. Rev. Lett.* **62**, 63 (1989).

6.29 D. L. Abernathy, S. G. J. Mochrie, D. M. Zehner, G. Grubel, and Doon Gibbs, *Phys. Rev. Lett.* **69**, 941 (1992).

SUBJECT INDEX